HUBBLE SPACE TELESCOPE

New Views of the Universe

BY MARK VOIT

HARRY N. ABRAMS, INC., PUBLISHERS
IN ASSOCIATION WITH THE SMITHSONIAN INSTITUTION
AND THE SPACE TELESCOPE SCIENCE INSTITUTE

CONTENTS

Foreword: **Hubble's Legacy** LAWRENCE M. SMALL / DANIEL S. GOLDIN   3

Prologue: **Edwin Hubble at the Threshold of the Cosmos**   4

Chapter 1:   **The Hubble Space Telescope**   6

Chapter 2:   **Planets**   13

Chapter 3:   **Stars**   23

Chapter 4:   **Galaxies**   36

Chapter 5:   **Universe**   49

Epilogue: **Beyond Hubble** ALAN DRESSLER   59

On Color   60

Acknowledgments   61

Credits   62

HUBBLE'S LEGACY

Humans have gazed at the sky for countless centuries, exploring the heavens. Until just a few hundred years ago, those explorations were limited to what we could see with our unaided eyes: the Sun, the Moon, a few planets, some ten thousand stars, and the indistinct smudges of three nearby galaxies. The invention of the telescope expanded and clarified that view a hundred fold, revolutionizing our understanding of the universe and how it works.

A second great revolution in our understanding of the universe is now under way. With the dawning of the space age, we are no longer limited to what we can see from beneath Earth's atmosphere. Space offers a much clearer vantage point for our telescopes, and the pictures being sent back tantalize us anew. Hardly a week goes by without news of some intriguing object in the cosmos, and many of these discoveries can be credited to scientists using NASA's Hubble Space Telescope.

The Space Telescope has changed our view of the universe and our place within it. Hubble's pictures provide us with a detailed view of the fascinating complexity and diversity of the universe, as well as its startling beauty.

Hubble has helped unlock some of the deepest mysteries of the cosmos during its first decade of operation, but other mysteries still beckon. An incredible future of discovery lies ahead, and even more powerful space telescopes are now being planned to further probe these puzzles and continue Hubble's legacy of exploration.

LAWRENCE M. SMALL

SECRETARY OF THE SMITHSONIAN INSTITUTION

DANIEL S. GOLDIN

NASA ADMINISTRATOR

EDWIN HUBBLE AT THE THRESHOLD OF THE COSMOS

Eighty years ago, the frontiers of the universe seemed to extend only a few thousand light-years beyond our solar system. We knew that our Sun was an ordinary star, drifting with its entourage of planets through a vast galaxy of stars, but many astronomers thought our Milky Way galaxy might be a lone island with nothing but blackness beyond its borders. Here and there we saw what appeared to be other swirling collections of stars, but no one had accurately measured their distances.

Then our perceptions of the universe changed dramatically, thanks largely to the efforts of American astronomer Edwin Powell Hubble (1889–1953). Hubble had at his disposal what was then the largest telescope in the world, the 100-inch Hooker telescope atop California's Mount Wilson. His first great breakthrough, in 1924, was to measure the distance to a spectacular spiral-shaped collection of stars in the constellation Andromeda, establishing it as a large galaxy lying far beyond the outskirts of our own. Suddenly the Milky Way no longer seemed so lonely.

Many more galaxies like Andromeda were out there, waiting to have their distances measured. Over the course of the next few years, Hubble's observations showed that the distances of galaxies are related to their speeds in a very special way: the farther a galaxy is, the faster it is moving away from us. All the galaxies in the universe were once much closer together, and in the distant future they will be much farther apart. In other words, the entire cosmos is expanding.

The implications of Hubble's discovery astounded the world. These galaxies could not have been drifting apart forever. Tracing their motions back through history shows that they must all have started on their journey at a particular moment in time, in an event we have come to call the Big Bang. Hubble's discovery therefore gives us a way to measure the age of the universe. By measuring the distances and motions of galaxies, we can determine how long they have been flying apart and thus date the moment of creation.

Edwin Hubble's work opened the door to our modern understanding of the universe as an expanding realm within which a constantly evolving story of creation is still playing itself out. However, the shortcomings of Hubble's own measurements prevented him from accurately age-dating the universe. A telescope with much finer vision than any existing ground-based telescope was needed. The telescope that finally completed his quest was named in his honor.

Edwin Hubble at the 100-inch Hooker Telescope

CHAPTER 1
THE HUBBLE SPACE TELESCOPE

n April 24, 1990, the space shuttle *Discovery* roared into space with historic cargo: the Hubble Space Telescope. Hubble's mission began a day later when the shuttle's robotic arm released it into Earth orbit. The world's most famous telescope has been snapping glorious pictures of the heavens ever since, as it orbits the Earth once every 97 minutes at an altitude of about 600 kilometers (370 miles).

Why expend such effort to place a large telescope in orbit? The primary reason is that the Earth's turbulent atmosphere disrupts the passage of light, blurring our view of the universe. To our naked eyes, the blurring is

A TELESCOPE IN THE SKY

very slight; we perceive it as the twinkling of stars. Through a powerful telescope, however, the blurring is much more evident, and it hindered progress in certain fields of astronomy for many years.

Another reason for placing Hubble above the Earth's protective blanket of air is that this blanket absorbs many kinds of radiation, preventing these rays from reaching earthbound telescopes. Ultraviolet light and many shades of infrared light do not reach the ground. From its perch above the skies, Hubble can detect these forms of radiation and discover clues to the workings of the cosmos that would otherwise remain hidden.

BUILT TO UPGRADE

Hubble, the largest visible-light observatory ever placed into space, is unique among astronomical satellites in that it improves with age. Anticipating that technology would advance during Hubble's long mission, NASA designed Hubble so that shuttle astronauts could upgrade it, installing newer, more capable instruments every few years.

During the first servicing mission in 1993, Hubble received a new camera, additional optics to compensate for flaws in the original mirror, and improved solar arrays. During the second mission in 1997, Hubble gained an infrared camera and a much improved spectrograph for analyzing the colors of cosmic light. The third servicing mission (above) in 1999 installed a new set of gyroscopes crucial to pointing the telescope.

Astronomers look forward to continued improvements during the next decade as future servicing missions continue to install newer and better instruments. If all goes well, Hubble will continue to beam back spectacular images until at least 2010.

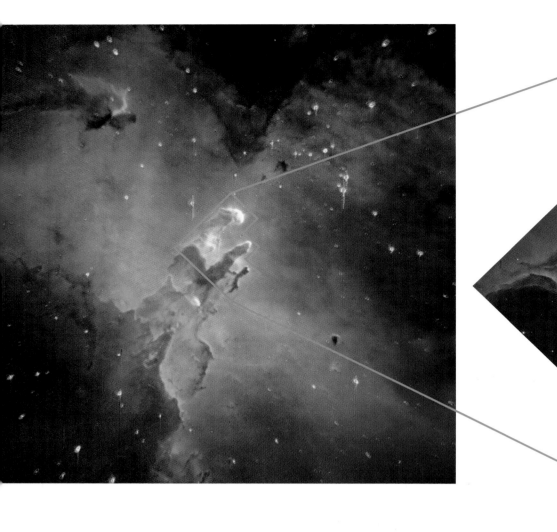

HUBBLE'S EXQUISITE VISION

The clarity of Hubble's vision is especially evident in highly detailed images such as that of the Eagle Nebula, above right. Stars are rapidly forming in this region of our galaxy, and their radiation heats the surrounding gas, causing it to glow. Three gnarled columns of dark, star-forming gas dominate the image, and Hubble's sharp eye resolves the wispy details and odd, fingerlike protrusions created as the hot radiation gradually erodes these columns away. To the upper left is an image showing how this same nebula looks through a ground-based telescope. The three gaseous columns stand out near the nebula's center, but

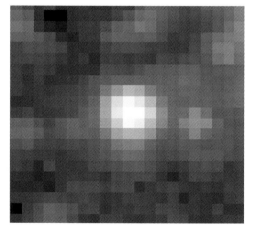

many of the finer structures are blurred beyond recognition.

Deeper appreciation of Hubble's power comes when we look at even tinier objects. To the left are two images of the same star, one taken by Hubble and one by a high-quality, ground-based telescope; both are shown to the same scale. In the ground-based image (top), the star appears as a blurry blob surrounded by fainter blobs. In the Hubble image (bottom), the star's image is much sharper, and the surrounding blobs turn out to be entire groups of stars that cannot be distinguished from the ground.

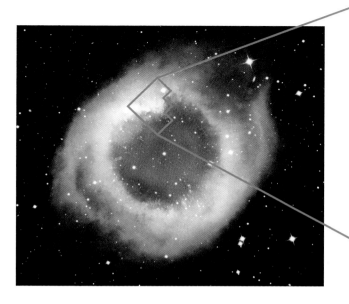

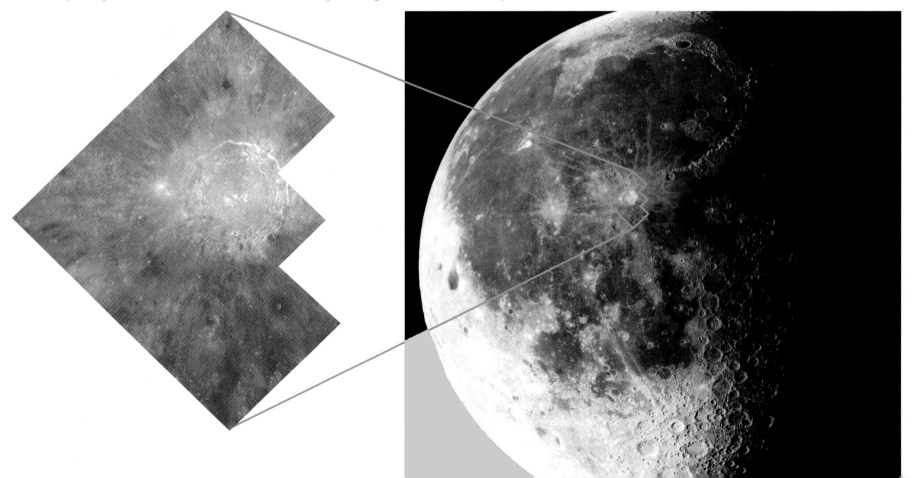

CLOSE-UPS OF THE COSMOS

Hubble's remarkable resolving power does entail a sacrifice. In order to take full advantage of Hubble's precise vision, we must train Hubble's cameras on extremely narrow regions of the sky. The photographs on this page show two examples. At the top is the Helix Nebula, a luminescent puff of cosmic gas from a dying star in the constellation Aquarius. If your eyes were sensitive enough to see this nebula from the Earth, its apparent size would be similar to that of the full Moon. Shown at the bottom is the Moon itself. In each case, a full-field Hubble photograph can reveal only a small portion of the whole.

The distinctive notched shape of each Hubble picture is governed by the design of Hubble's main visible-light camera. Four light-sensitive silicon chips, similar to those you might find in a video camera, are arranged in a four-square pattern. However, one of these chips, designed to deliver ultra-precise imaging, is trained on an even smaller piece of the sky than the others. Three-quarters of its quadrant goes unobserved, leading to the unusual footprint of Hubble images.

Next time you notice the full Moon, try covering it up with the tip of your pinkie. You will probably achieve this goal easily, even when holding your hand at arm's length. Most people are surprised to find that the Moon's image is really this small, yet Hubble's field of view is smaller still. To visualize Hubble's pinhole view for yourself, you might want to draw a small dot on your pinkie's fingernail. Now hold your hand at arm's length. Each of the Hubble photos displayed in this book represent a region of the sky roughly the same apparent size as that dot. Hubble can't see much at any one time, but it is a close-up camera without parallel.

VIEWING THROUGH SPACE ...AND TIME

Because light travels at a finite speed, we are always seeing astronomical objects as they were sometime in the past. Much of the light Hubble records has been on its way to Earth for quite a while. The graphic to the right shows just how far we are looking back in time when we observe various kinds of celestial bodies.

PLANETS: The news Hubble gathers on our solar system is relatively current. Light reaches us from the nearest planets, Venus and Mars, in only a few minutes. Light from Jupiter and Saturn takes a little longer, about an hour, and light from the farthest planets, Pluto and Neptune, takes about four hours to get here. We are therefore seeing our solar system as it was during the past day.

STARS: By contrast, the news from the stars in our galaxy is much older. Distances to stars and nebulae in the Milky Way range from a few light-years to tens of thousands of light-years. Thus, we are seeing them as they were sometime during human history, from several years to several thousand years in the past.

GALAXIES: Light from all but the very nearest galaxies predates the appearance of humans here on Earth. The Andromeda galaxy, the nearest large galaxy to our own, lies about 2.5 million light-years distant, and Hubble has measured precise distances to galaxies as far as 100 million light-years away. The light from those galaxies left when dinosaurs roamed our planet.

UNIVERSE: Hubble's deepest soundings of the universe detect light that began its journey long before the Earth even existed. Current estimates for the age of the universe range from about twelve to sixteen billion years, and this time-span limits how much of the universe we can observe. Light from objects more than twenty billion light-years away, if there are any, would not yet have had time to reach us. The most distant objects that Hubble can detect are galaxies from when the universe was about a billion years old. When we observe them we are seeing back over 90 percent of the way to the beginning of time. These galaxies are quite young, but the light from them is ancient.

LIGHT TRAVEL TIME TO EARTH

PLANETS

STARS

GALAXIES

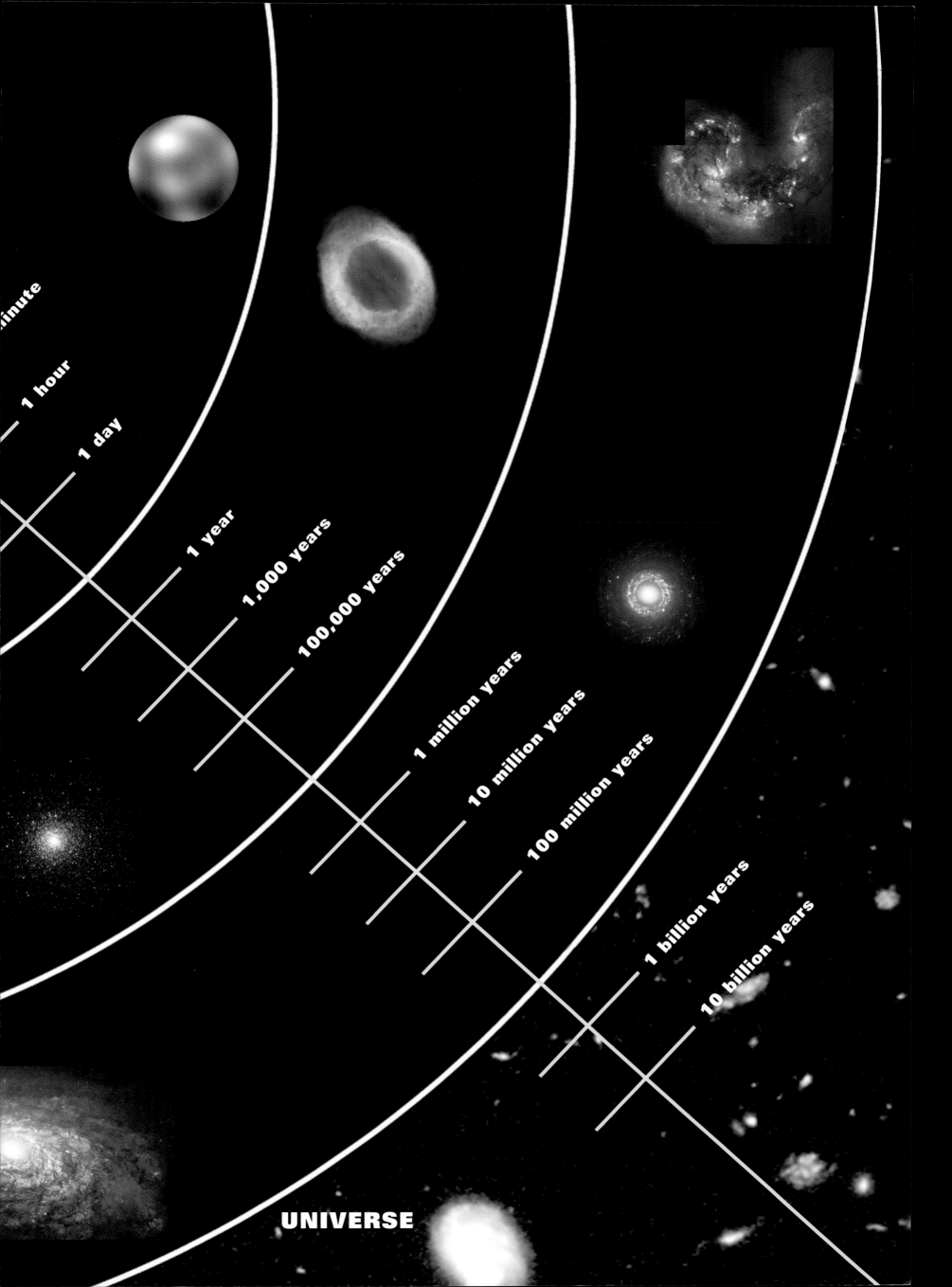

inute

1 hour

1 day

1 year

1,000 years

100,000 years

1 million years

10 million years

100 million years

1 billion years

10 billion years

UNIVERSE

CHAPTER 2
PLANETS

A HUBBLE ALBUM OF PLANETS

Hubble has observed seven of the nine planets orbiting the Sun: Venus, Mars, Jupiter, Saturn, Uranus, Neptune, and Pluto. Earth and Mercury are missing from this family album because Hubble cannot observe them. Earth is too close for Hubble to see more than a tiny piece of it at any one time. Mercury, the nearest planet to the Sun, is too hazardous to observe because a small mistake in aiming the telescope might point Hubble directly at the Sun, permanently blinding its cameras.

MARS

Russet Mars is a frequent Hubble target. This visible-light picture shows the planet during Martian summer in 1997. Its northern polar cap of white carbon-dioxide ice (top) has receded since the winter, exposing the darker brown sand dunes beneath.

VENUS

This Hubble picture of Venus was taken in ultraviolet light. Perpetually-cloudy Venus is seen here close to half phase because Earth, Sun, and Venus formed a near ninety-degree angle at the time. This arrangement is the safest for Hubble observations of Venus, because it maximizes the separation of Venus from the Sun.

JUPITER

This visible-light Hubble image of Jupiter, the largest planet, also features Io, one of Jupiter's largest moons. Io was passing in front of the planet when this picture was taken, and the black dot to the left of center is Io's shadow. If you look carefully, you can pick out Io itself against Jupiter's cloud tops (just to the right of the shadow).

SATURN

Saturn, the second-largest planet, is not nearly as psychedelic as it appears here. The colors in this image, taken in infrared light, were chosen to bring out Saturn's atmospheric features.

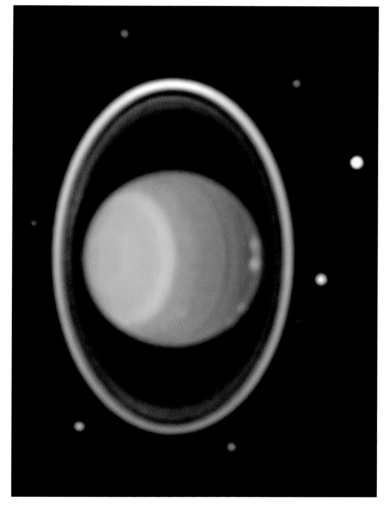

NEPTUNE

Neptune is shown here in visible light, but its colors have been enhanced to show storm activity in its upper atmosphere.

URANUS

Like Hubble's view of Saturn, this image of Uranus shows the planet and some of its moons in infrared light. Its rings are clearly visible, and their orientation is unusual. Unlike most planets, whose poles point perpendicular to the Sun, Uranus and its rings rotate around an axis that points directly at the Sun twice each orbit. As a result nighttime at the poles lasts forty-two Earth years.

PLUTO

Hubble's sharp vision has given us the only pictures that reveal Pluto's surface features. This computer-processed version of ultraviolet-light images from Hubble show a world whose icy surface is mottled with light and dark patches.

A COSMIC CATASTROPHE

Hubble's first servicing mission at the end of 1993 arrived just in time to prepare the telescope for one of the most remarkable events ever witnessed in our solar system. In June 1994, some two dozen pieces of comet Shoemaker-Levy 9 slammed one by one into Jupiter's surface. A previous close encounter with Jupiter had torn the comet to pieces before sending it on a fateful boomerang course. The composite Hubble image on the facing page shows the fractured comet alongside its ultimate destination.

Each fragment of comet Shoemaker-Levy 9 was assigned a letter; the Hubble image below shows fragments A through W. The largest fragments measured roughly a kilometer (0.6 mile) across and weighed almost a billion tons. Each icy chunk imparted an energy equivalent to a million hydrogen bombs as it plunged below Jupiter's cloudtops, kicking up a plume of hot gas thousands of miles above Jupiter's surface.

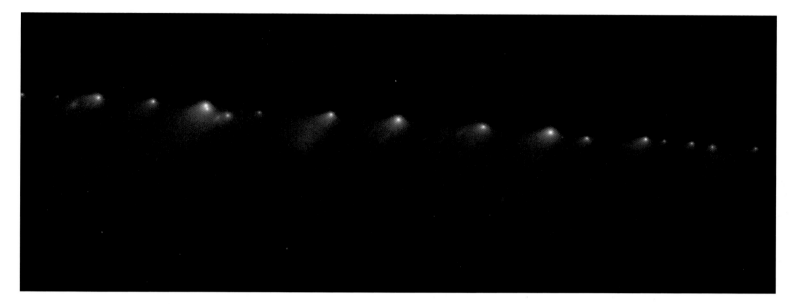

JUPITER GETS A BLACK EYE

The time-lapse photo sequence to the right depicts the aftermath of a single impact. In the top image we see the entry plume of fragment G, one of the largest, billowing above Jupiter's edge. The location of the impact site, slightly around the far side of Jupiter, prevented us from seeing the actual collision; astronomers had to wait a few minutes for the plume to rise high enough to observe. Jupiter's rotation brought the impact site into view ninety minutes later. The second image shows the eyelike ring of dusty debris scattered across Jupiter's surface at that time. Days later the debris still remained suspended in Jupiter's clouds, covering a region larger than the Earth. The final two images show the impact site three and five days after the initial catastrophe.

Astronomers estimate that impacts like this now occur only once every few thousand years, but they were much more common in the past. Our solar system's planets appear to have formed from a gaseous disk of dusty debris that orbited the young Sun some 4.5 billion years ago. Over the course of a few hundred million years, this debris clumped into rocks, boulders, and mountain-sized lumps that ultimately collected into planets. Like its planetary companions, the young Earth withstood a steady rain of rock and ice from outer space as it slowly assembled itself. Events like cometary impacts thus echo the solar system's formation.

Most of the comets and asteroids originally orbiting the Sun either plunged into planets or were ejected from the solar system long ago. The few that remain occasionally meet fates like that of comet Shoemaker-Levy 9, and the Earth itself is sometimes the target. A similar collision sixty-five million years ago probably sent the dinosaurs to their extinction, and someday our planet will experience yet another disastrous impact.

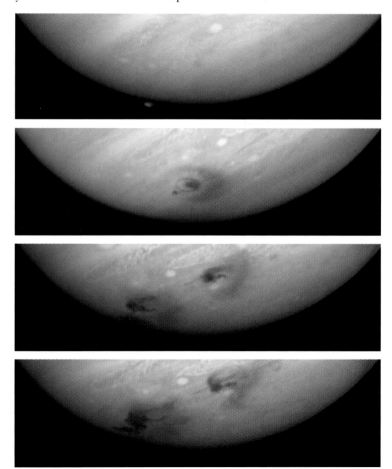

MARTIAN WEATHER

Many NASA spacecraft have explored the solar system during the last few decades. Most have been probes dispatched to observe individual planets in detail for limited periods of time. Hubble's unique contribution to this effort is its ability to observe just about any planet whenever something notable is happening.

The Hubble images of Mars below were taken in sup-port of the 1997 landing of Mars Pathfinder. The June 27 image (left) shows an unsettled Mars. Whitish clouds cover large portions of the northern hemisphere, and a yellowish dust storm rages near the equator at the left of the image. Two weeks later, on July 9, Mars had become more hospitable, as is evident in the photo on the right. The clouds have largely dissipated and the planet's equatorial features are no longer immersed in yellowish haze.

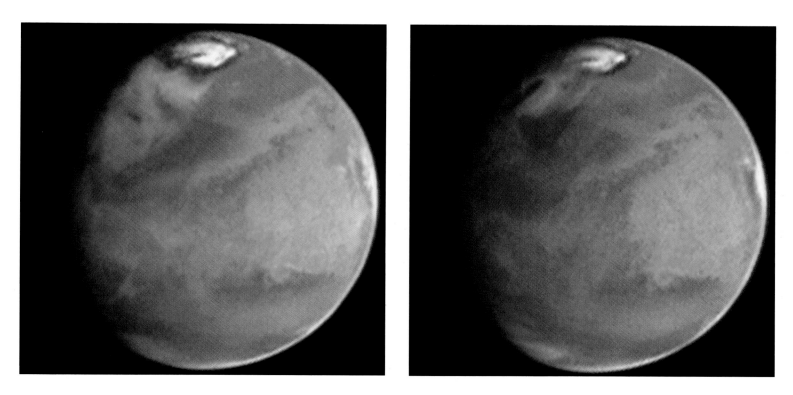

A CYCLONE ON MARS

Hubble's monitoring of Mars continues to reveal other weather curiosities. In April 1999, Hubble spotted the cyclonic cloud formation swirling near Mars' north pole seen on the facing page. Because this picture was taken during the height of summer in Mars' northern hemisphere, the north polar ice cap is relatively tiny. Much of the seasonal carbon-dioxide ice that covers the poles during Martian winter had temporarily returned to the atmosphere, revealing the deeper layers of permanent water ice.

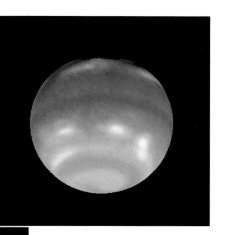

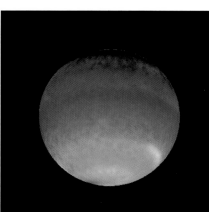

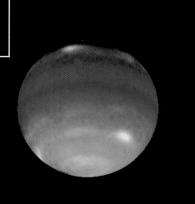

NEPTUNIAN WEATHER

These pictures track storm activity on Neptune. The two on the left, showing different sides of the planet, were taken on August 11, 1998. Several whitish blotches represent disturbances in Neptune's upper atmosphere. The two pictures on the right show that by August 13, these storms had subsided.

SOLAR SYSTEM SURVEILLANCE

Hubble's surveillance of the solar system is not always seeking changes in the weather. The ultraviolet-light image on the facing page features Saturn's equivalent of Earth's northern lights. Charged particles streaming from the Sun are stimulating atmospheric atoms near Saturn's poles to glow in response. The resulting auroras resemble those seen in Earth's polar regions.

IO VOLCANO

In the picture of Jupiter and Io to the right Hubble has caught a sight rare to behold beyond Earth: a volcano in the act of erupting. Ever since the Voyager 1 space probe flew past Io in March 1979, we have known that this moon of Jupiter is volcanically active. But after Voyager 1's visit and that of Voyager 2 a few months later, Io's eruptions went unobserved for two decades. Then, on July 24, 1996, Hubble was watching when the Pele volcano on Io blew its stack. The color inset shows a close-up view of Io with Pele's plume visible above the moon's edge at an orientation of about four o'clock.

JUPITER'S AURORAS

Hubble has detected auroras on Jupiter as well as on Saturn. The main two auroral rings in the picture to the left were stimulated by solar activity, like auroras on other planets. On closer inspection, however, one can see two smaller dots of light: one in the north and one in the south, slightly farther from the poles than the auroral rings. The object responsible for these smaller auroras is Io, whose volcanoes spew charged particles that eventually reach Jupiter's atmosphere.

CHAPTER 3
STARS

SNAPSHOTS OF STELLAR LIVES

To our human eyes, stars appear steady and unchanging, the very definition of eternity. Yet stars progress from birth to death as surely as we humans do. They are made mostly of hydrogen gas and generate energy primarily by fusing the nuclei of hydrogen atoms to form helium. Nuclear reactions like these supply enough energy to keep a star shining for millions to billions of years, but eventually the nuclear fuel runs low. The consequences of this terminal energy crisis are unavoidable. With no more energy available, the star is doomed to fade into blackness. Because a star's lifetime vastly exceeds that of a human, we cannot see the whole of a star's life. Instead we must piece together its history from snapshots of different stars at different points in their lives, just as we might reconstruct the life history of a typical human from snapshots in a family album. Hubble's spectacular images of stars have greatly aided this effort because they reveal details only hinted at in images taken from the ground. The Hubble images on these pages illustrate how stars progress from a cloudy birth in a dark pocket of interstellar hydrogen gas to glorious death in a fantastic celestial light show.

CRADLE OF CREATION

The portion of the Eagle Nebula shown above marks a region in the Milky Way where stars are currently forming. Inside this large, black column of dusty gas, gravity is drawing hydrogen molecules together into tight clumps that will soon form the cores of newborn stars. Meanwhile, ultraviolet radiation from hot, young stars just outside the borders of the picture scorches the surface of this dark, cloudy column, boiling off its gases and causing it to glow with striking iridescence.

AT BIRTH

Stars form because gravity inexorably pulls interstellar gas molecules together into dense clumps. As these molecules accumulate, the clumps begin to spin and flatten into star-forming disks like this one. Matter spiraling inward through such a disk gradually collects at the disk's center where the star is being assembled.

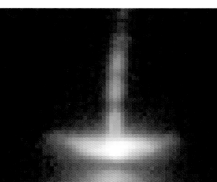

In this image, light from the invisible infant star illuminates the top and bottom surfaces of the disk, depicted here as two red layers sandwiching the dark interior. High-speed jets of matter (green), created by a process that remains mysterious, simultaneously shoot out along the disk's rotation axis.

YOUNG STAR CLUSTER

Large interstellar clouds spawn stars in clusters that continue to shine long after the cloud's dense gases have dissipated. This Hubble image shows a star cluster shortly after birth. Dark gaseous columns, reminiscent of the Eagle Nebula, still linger at the outskirts of the cluster, their skins aglow, as the searing radiation from the cluster's hot, blue stars erodes them into nothingness.

NEARING THE END

Betelgeuse, the bright red "shoulder star" in the constellation Orion, is a supergiant star in its last phase of life. All stars grow large and red near the ends of their lives, but this particularly massive star has grown larger and redder than most. If it were placed at the Sun's location, its bloated surface would easily engulf the Earth's orbit, and its flames would lap at Jupiter. Yet, even the largest stars look like mere points of light when viewed through ground-based telescopes. This Hubble image of Betelgeuse, taken in ultraviolet light, was the first telescopic image to reveal surface features on a star other than our own Sun.

COSMIC RECYCLING

Hubble has caught the spectacular end of this doomed star, known as NGC 7027. During its life, the star has burned hydrogen into helium and helium into carbon, oxygen, and heavier elements. Now it is returning these elements to space, a process that will take thousands of years. Gravity, acting on the gases between the stars, will ultimately reincorporate these elements into new clouds, stars, planets, and perhaps even living beings like ourselves.

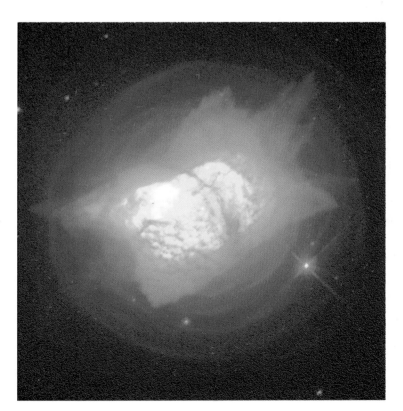

STAR BIRTH IN ORION

Fifteen hundred light-years distant, visible with binoculars just below Orion's three-star belt, lies the Orion Nebula, one of the nearest places where stars are actively forming. Fifteen separate Hubble images of this nebula were pieced together to construct the mosaic on the facing page. Ultraviolet radiation pouring off the prominent group of four young stars near the center heats and ionizes the surrounding gas, causing it to glow. The dark blotches at the nebula's margins are places where fingers of dusty gas protruding from the large cloud in the background are wrapping around the hot blister created by the stars' scalding radiation.

PLANETS IN THE MAKING

Hubble has found brand-new stars silhouetted against the Orion Nebula that may soon host new planetary systems. The photos above show newborn stars—the red dots—girdled by dark, dusty disks of orbiting gas that block the light from the nebula behind. At the right we are seeing one of these disks edge-on. Some five billion years ago our own solar system probably looked similar. Then, over the course of a few hundred million years, the dusty material clumped into the nine planets orbiting the Sun today. Many disks such as these, scattered throughout our galaxy, are currently forming the planets of the future.

A SMOKING GUN

More often than not, star-forming disks are difficult to see because they are buried deeply in clouds of dark gas. However, some of these hidden infant stars still manage to give themselves away. Jets of matter streaming from young stars can drill their way through the surrounding cloud, leaving behind a luminescent wake. The Hubble picture to the left shows one such jet, shooting like a blast from a fire hose through interstellar space. The source of the jet is buried in dark gas at the lower left of the picture.

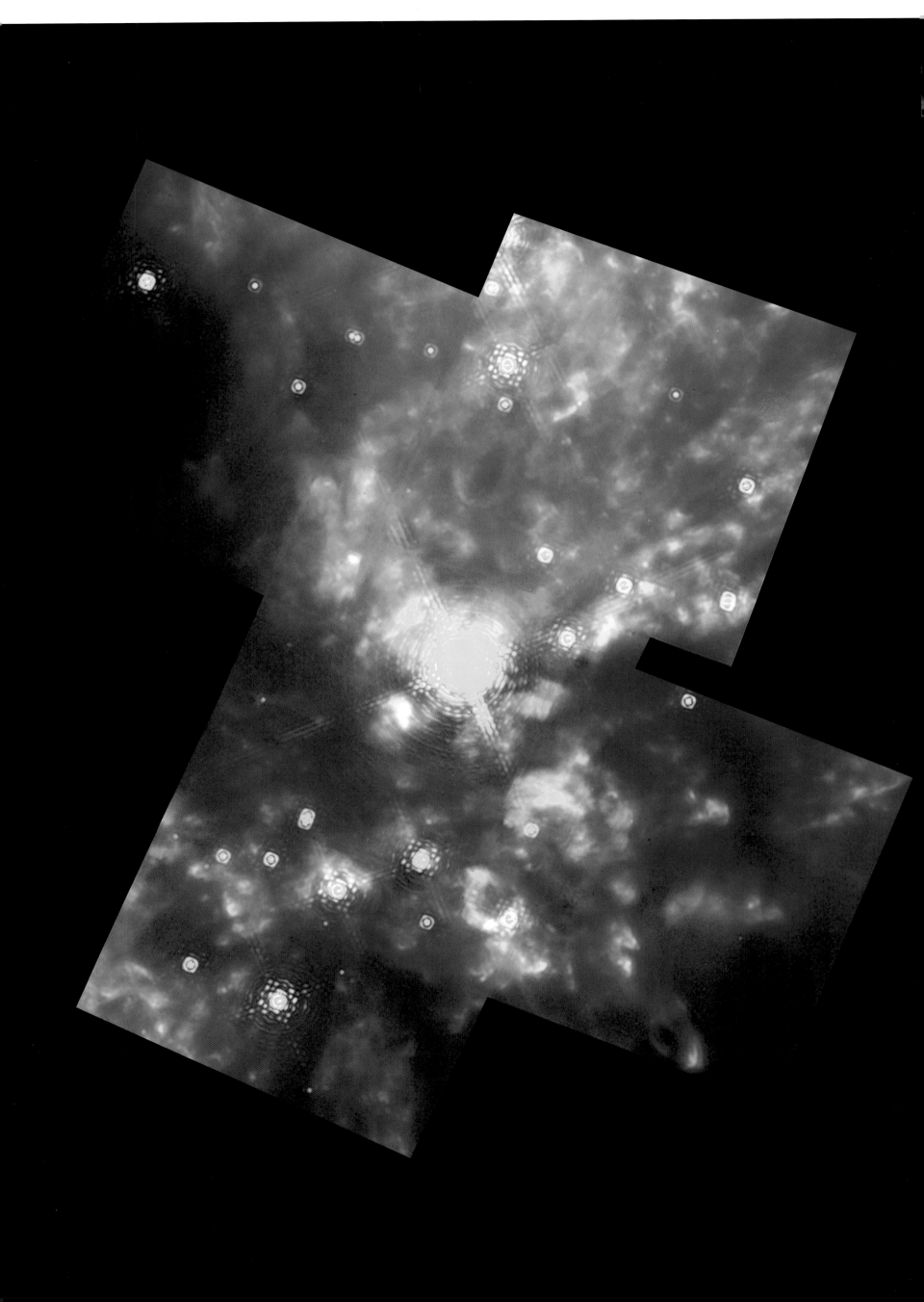

HIDDEN STARS

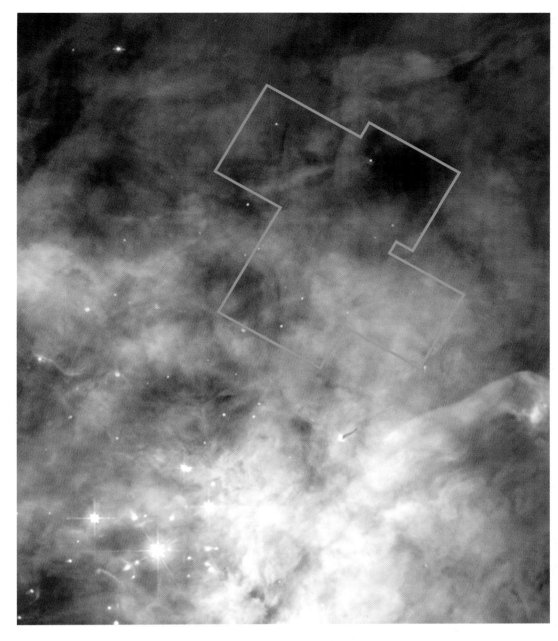

The smokescreen of dusty gases pervading our Milky Way galaxy hides much of its visible light from direct view, particularly in star-forming regions such as the Orion Nebula. That is one reason why NASA upgraded Hubble with an infrared camera in 1997. Infrared light penetrates these dusty shrouds, revealing the goings-on deep within our galaxy's clouds.

The Hubble image on the facing page shows a portion of the Orion Nebula in infrared light. Yellow tones correspond to infrared starlight, and blue tones depict infrared light emitted by the cloud's hydrogen molecules. At the center of this image shines a bright star that can't be seen in the visible-light image to the right. The blue outline on the visible-light image defines the boundaries of the infrared-light image. Only a few stars appear within the outline and nothing at all appears near the outline's center.

SEEING THE INVISIBLE

Hubble's infrared vision is handy for more than just peering through interstellar smokescreens. Certain objects, for example the Egg Nebula in the pair of images below, look completely different in infrared light. This nebula surrounds a dying star that is casting off its outer layers of gas. In optical light (left) we see four mysterious beams penetrating like searchlights through fragmented, multi-layered shells of gas. In the infrared (right), we see scattered starlight (blue) and molecular hydrogen emissions (red) aligned along those searchlight beams, but we also see a strange ring of hydrogen molecules extending perpendicular to the beams.

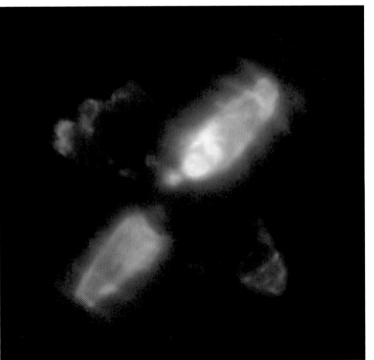

A MULTICOLORED MYRIAD OF STARS

As you look at the picture on the facing page, you are gazing along a line of sight passing not far from the Milky Way's center, toward the general direction of the constellation Sagittarius. This image illustrates Hubble's ability to distinguish individual stars in crowded fields of the sky. Many of the tiny stars populating this densely-packed region would blend into a hazy blur in a photo taken from the ground.

The image illustrates another point: that stars range widely in color, size, and brightness. The most common stars are tiny, red, and dim; they are also the longest lived, burning for several tens of billions of years. Also common are yellow stars like our Sun, with shorter lifetimes of ten billion years or so. Less common white stars generate even more light than the Sun but live even shorter lives. Blue stars, the rarest and brightest of all, burn for only a few million years. Color, brightness, and lifetime are so closely related because they all depend on a star's mass. Crushing pressure within the core of an unusually heavy star causes it to burn furiously, rapidly exhausting its fuel, while its extraordinarily hot surface radiates enormous amounts of brilliant blue light. The core pressure within a less massive star is not so severe. It consumes its fuel more modestly, and its cooler surface radiates a redder shade of light.

The most prominent stars in the image seen here are exceptions to this general trend. They are orange and red stars that have exhausted almost all their fuel. The common fate of all stars, whatever their color, is to grow larger, brighter, and redder at the ends of their lives, as their cores strain to squeeze out the last few measures of nuclear energy.

AGE-DATING THE COSMOS

Large clouds of interstellar gas bring forth clusters of stars whose colors span the spectrum. Because a cluster's short-lived blue stars die off long before the white and yellow stars, the blueness of a star cluster tells us its age: clusters with many blue stars are young and those with few blue stars are old. The infant cluster at the upper left, whose light is dominated by blue stars, is only a few million years old. In the middle is an adolescent cluster, about forty million years of age. Some of its original blue stars are still present, but others have ripened into large red stars approaching death. The reddish cluster at the right is an ancient relic of the early universe some twelve to sixteen billion years old. Its original blue stars died off eons ago.

Astronomers have been age-dating clusters in our own galaxy for decades, but Hubble's superior vision allows us to age-date star clusters well beyond our galaxy's borders. Hubble's clarity also helps us age-date nearby star clusters more precisely. Age estimates for the oldest clusters range from about twelve to sixteen billion years. The universe itself must be at least this old but probably not much older.

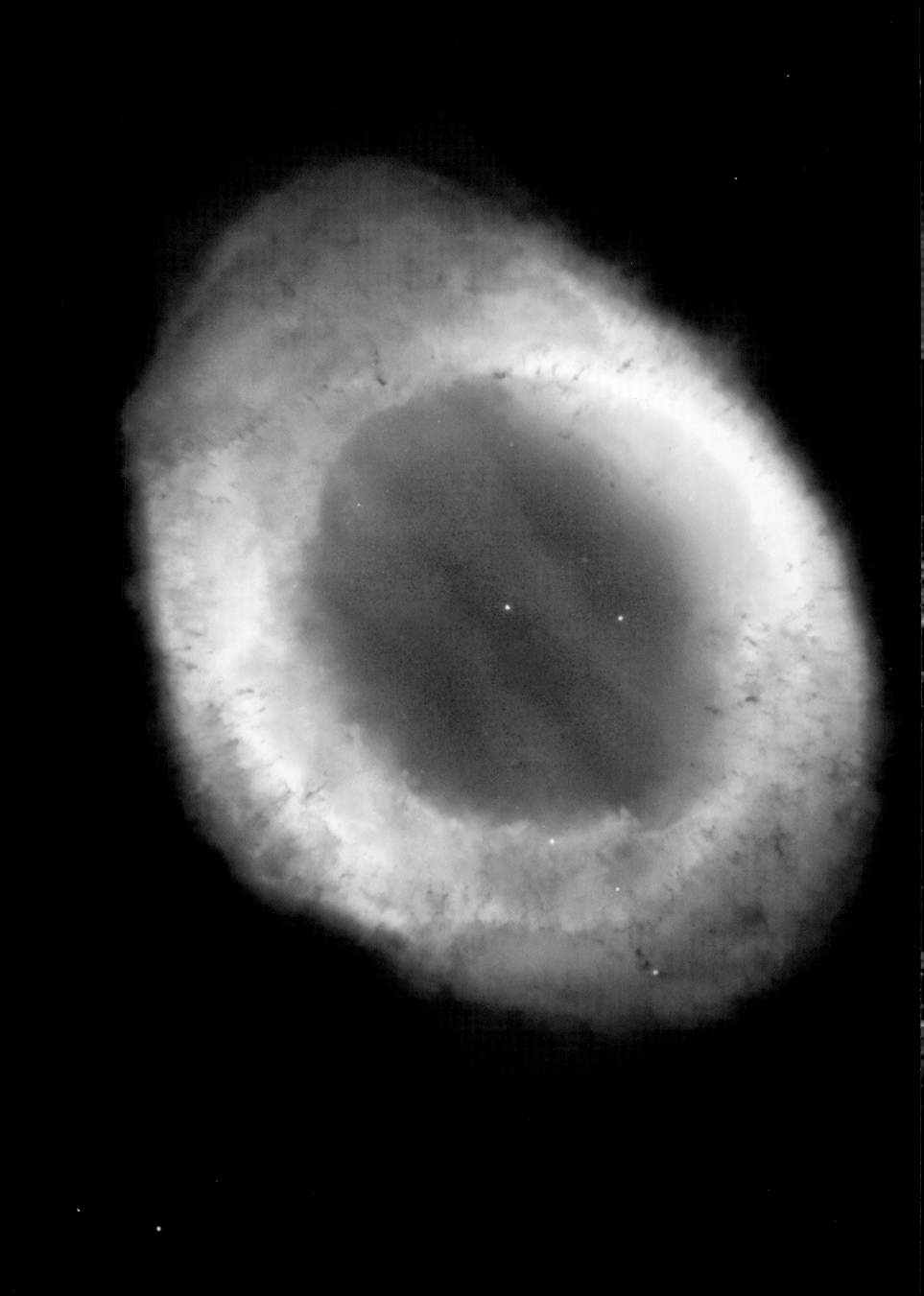

IMAGES OF STAR DEATH

Old stars do not fade quietly away. Instead they dramatically blow off their outer layers, creating spectacular celestial fireworks. All the pictures on these two pages show stars expelling their dying breaths. Astronomers call these fluorescent displays "planetary nebulae."

The Ring Nebula on the facing page is a favorite of amateur astronomers. We are looking down a barrel-shaped puff of ejected gas energized by the white-hot ember at the nebula's center. This ember was once the star's nuclear furnace. No longer able to burn on its own, it is doomed to become increasingly cooler and dimmer. In a few thousand years the bright surrounding gases will fade and dissipate, leaving only the waning ember—a nondescript white dot in the cosmos.

The story is similar for the other three nebulae shown below, even though their manifestations, revealed by Hubble in unparalleled detail, look so different. The Hourglass Nebula to the lower right and the Twin-Jet Nebula on the bottom of the page may well have dense equatorial belts of matter that direct the glowing gases out along their poles. The intricate symmetry of the Cat's Eye Nebula below left suggests that it may be ejecting its gases like a twirling lawn sprinkler.

When we look at nebulae like these, we are looking far into the future of our own solar system. Roughly five billion years from now the Sun will exhaust its nuclear fuel, grow red and large, and blow away its outer layers in a similar display. Perhaps by then humanity will have figured out how to watch the fantastic show from a safe distance.

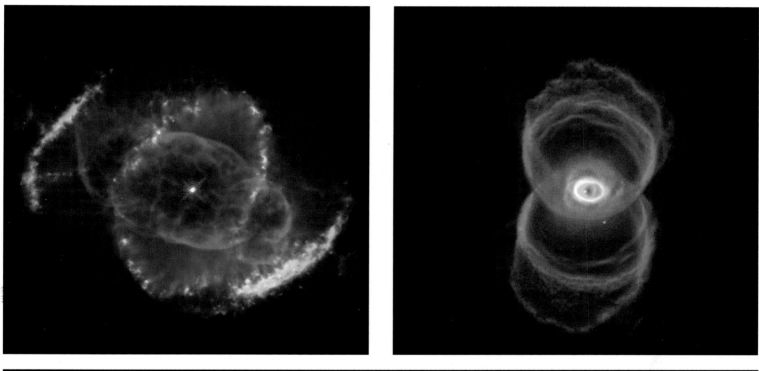

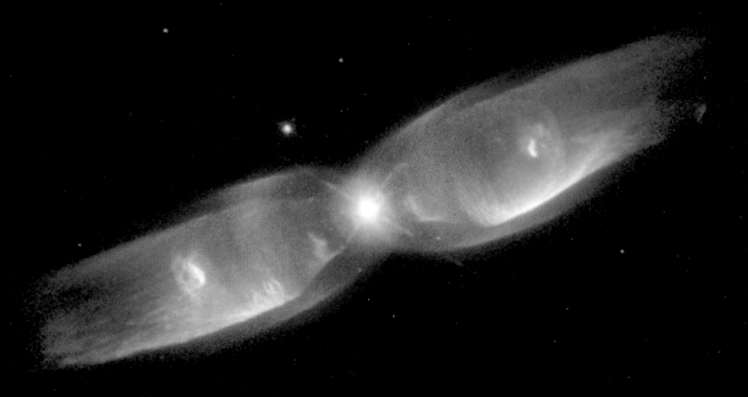

AN EXPLOSIVE DEMISE

The brightest and bluest of stars—those that begin their lives more than ten times heavier than the Sun—catastrophically explode when they die, shining brilliantly for a few brief weeks as supernovae. Three years before the space shuttle *Discovery* carried Hubble aloft, watchers of the southern skies were treated to a rare celestial spectacle, the first supernova visible to the naked eye in almost four centuries. This explosion, christened Supernova 1987A because it was the first supernova discovered that year, occurred about 160,000 light-years away in the Large Magellanic Cloud, a small companion galaxy to the Milky Way. Ground-based photos (right) show the neighborhood of SN 1987A before and after the explosion.

What remains of this supernova can be seen in the Hubble image on the facing page. Near the center is a dot surrounded by three rings that seem to intersect. The dot is the remnant of the supernova; the rings are part of a gaseous structure that predates the supernova. The rings' underlying shape, origin unknown, is probably like an

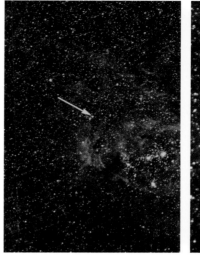

hourglass seen from an angle; the outer rings trace the top and bottom edges of the hourglass, while the inner ring traces its waist. Light from the supernova's flash energized this surrounding gas, which continues to glow more than a decade later.

Hubble will be keeping a close eye on the inner ring for the next several years. Debris from the supernova has been hurtling toward this ring ever since the original explosion and will be colliding with it during the next decade. When the collision occurs, the ring will glow far more intensely, perhaps revealing clues about what it was doing there in the first place.

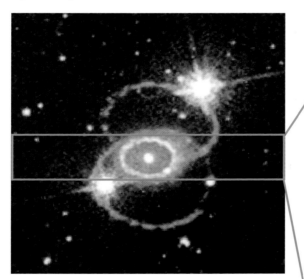

DISSECTING LIGHT

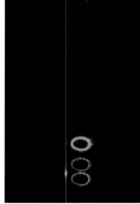

Hubble's fabulous pictures have become world famous, but its other less-publicized capabilities are just as important to science. Much of Hubble's time is devoted to analyzing the universe with instruments called spectrographs that dissect light like a prism, breaking it into various colors. Careful studies of this type can reveal an object's velocity, temperature, composition, and more.

Here we see how Hubble has broken the light from Supernova 1987A's central ring into several isolated colors, which are visible in the spectrum on the right. The ring essentially glows like a neon sign, with particular atoms producing very specific hues. Hydrogen atoms are responsible for the

blue image of the ring at the bottom of the lower band and for the middle of the three red-orange images at the bottom of the upper band; ionized nitrogen is responsible for the other two red-orange images; ionized sulfur generates the dim red images at the top, and doubly-ionized oxygen produces the two greenish rings. Light from the star to the lower left of the ring is also entering the spectrograph, creating the continuous stripe running to the left of the rings, similar to the spectrum of an incandescent light bulb. Various applications of this light-splitting technique have made it possible to measure the compositions and physical conditions of objects throughout the cosmos.

SWIRLING SPIRALS OF STARS

Galaxies are vast collections of stars and gas clouds, tens of thousands of light-years in size, all bound together by gravity. The galaxy known as NGC 4414, pictured above, is a typical spiral galaxy. Light from NGC 4414 began its journey to Earth around sixty-two million years ago, but even at this great distance, NGC 4414 is still considered a relatively nearby galaxy.

If we could gaze upon our own Milky Way galaxy from outside its borders, it would look similar to NGC 4414. We would see a bright bulge of stars at its center surrounded by a thin, flat, gas-filled disk decorated with sweeping spiral arms. In this Hubble picture of NGC 4414, the disk is tilted to our line of sight. The spiral galaxy NGC 7742, pictured right, is more face-on. With its yellowish bulge and whitish disk, this galaxy looks somewhat like a fried egg.

Spiral galaxies might seem like giant pinwheels, but if we could watch them over a long period of time, they would appear more like whirlpools. The inner regions rotate faster than the outer parts, and the spiral arms swirl like curling waves through the disk. You would need to be patient to see all this, however. A galaxy like the Milky Way completes a rotation only once every 200 million years or so.

COSMIC DIVERSITY

The universe's galaxies come in a wide variety of shapes, colors, and sizes. Most of the larger ones are spiral galaxies like our own. The galaxy group shown above contains three picturesque examples: one edge-on (bottom), one face-on (center), and one somewhat tilted (upper left). Many other large galaxies in the universe are more rounded, like the yellowish elliptical galaxy on the right side of the group. Occasionally we see rarer beasts like the peculiar galaxy pictured to the right. This "polar ring" galaxy apparently resulted from a collision between two more normal galaxies leaving an edge-on disk of stars and gas skewered in the center by a smaller, more rounded galaxy.

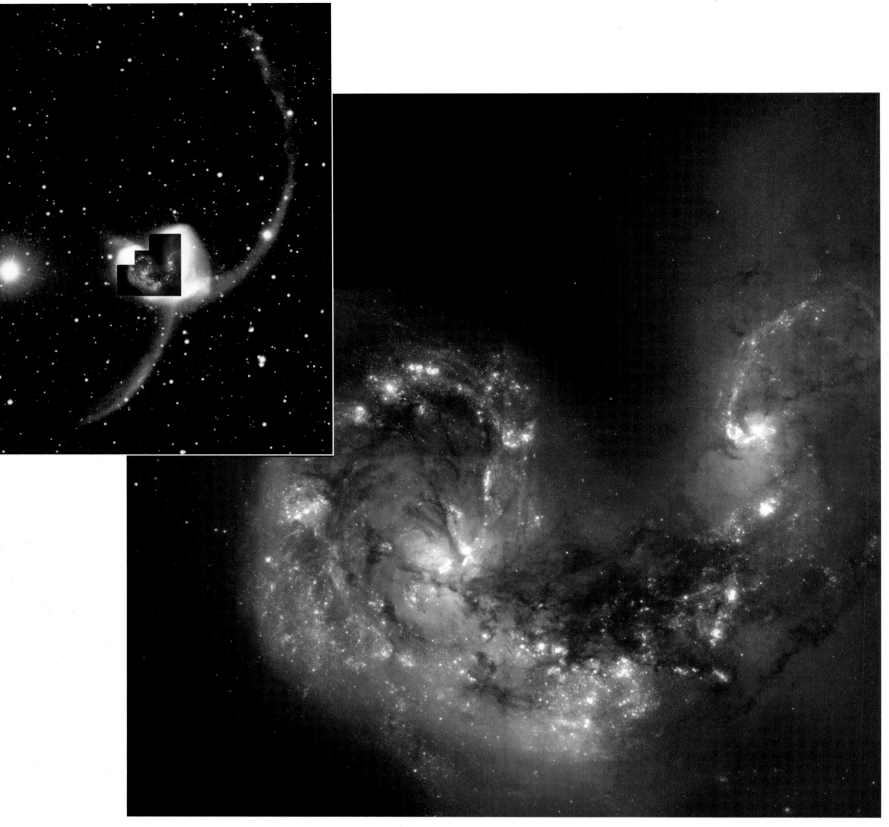

WHEN GALAXIES COLLIDE

Most galaxies spin through space in peaceful isolation from one another. Not so for the ill-fated "Antennae" galaxies pictured above. This pair of spiral galaxies is currently colliding. Two antennalike streamers of stars, stripped from these galaxies and flung into intergalactic space by gravitational forces, now extend many thousands of light-years from the two colliding disks. Gravitational forces will continue to tear apart these galaxies over the next several hundred million years, randomizing the trajectories of their stars and ultimately transforming the whole tumultuous mixture into a ball-shaped elliptical galaxy.

The Hubble picture of the Antennae, shown as a color inset in the wider angle ground-based view on the left, reveals the chaos at the current point of impact. Throughout this region we see dark patches—the signature of dusty gas clouds stirred up by the collision—interspersed with blue clusters of young stars. The sheer number of blue star clusters is astounding; the collision appears to have spawned a burst of new star formation. Hubble's measurements of these clusters' colors indicate that this particular population explosion began about a hundred million years ago, probably about the time the two disks came into contact.

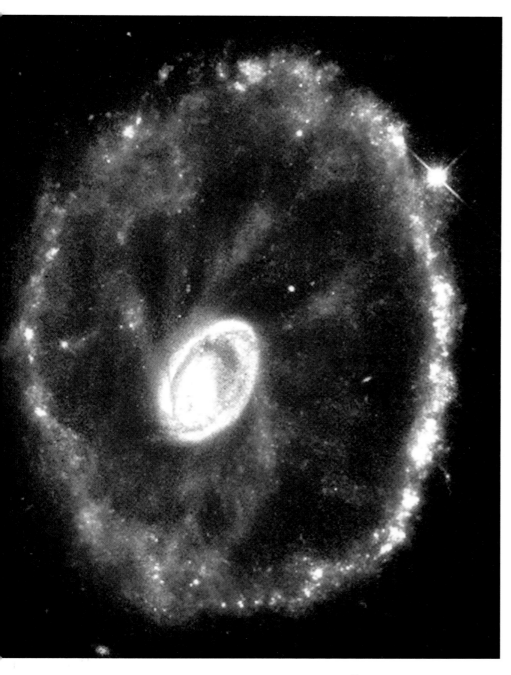

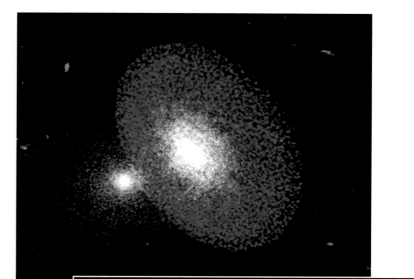

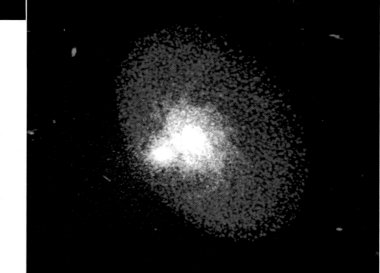

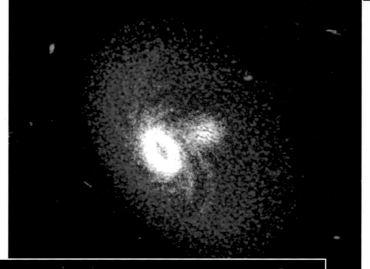

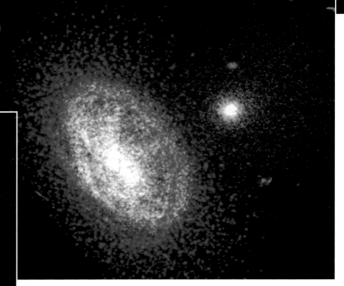

HITTING THE BULL'S EYE

Pictured above is the very unusual "Cartwheel Galaxy." Apparently a smaller galaxy has scored a direct hit on the Cartwheel's center. The five-frame sequence right depicts a supercomputer simulation that attempts to reconstruct the Cartwheel's past. The smaller reddish galaxy imparts a gravitational jolt as it passes through the center of the larger galaxy's disk. In response, the larger galaxy's stars and gas first move inward, then rebound, causing an expanding ring of stars and gas to propagate outward. Because the simulated galaxy ends up looking so much like the Cartwheel, astronomers suspect that the Cartwheel's history must have been quite similar.

DISASTER ZONE

The spectacular Hubble picture on the facing page shows the core of Centaurus A, a neighbor of the Milky Way lying about ten million light years away. The wider-angle composite photo of the Hubble image set into a ground-based photo below reveals Centaurus A to be an elliptical galaxy with a dramatic streak of dusty clouds obscuring its center, apparently the aftermath of a collision between galaxies. This streak of dust is all that remains of the unfortunate galaxy—or galaxies—torn apart by the violent encounter.

. . . WITH A MONSTER IN THE MIDDLE

Astronomers have long known that Centaurus A is unusual in other respects. For example, vast streams of particles shoot from the galaxy's center at nearly the speed of light, strong evidence that Centaurus A harbors a massive black hole.

Hubble pictures taken in dust-penetrating infrared light indeed show something peculiar behind the dusty gas that shrouds this galaxy's core. The bright spot in the infrared image to the right marks the presumed location of the black hole. However, the light we are seeing cannot be from the black hole itself. A black hole is completely black because not even light can escape its immense gravitational pull. The infrared light probably comes from a dense knot of stars drawn into tight orbits around the black hole.

BLACK HOLE HUNTING

Hubble is one of the best tools for black hole hunting available to astronomers. Because we cannot see black holes themselves, we look instead for the telltale motions of stars and gas clouds in their vicinity. Objects near a black hole will appear to be moving under the influence of an immense but unseen source of gravity at speeds that reflect the black hole's mass.

The Hubble image on the facing page shows the environs of the first suspected black hole to be verified by Hubble. It lurks at the center of the elliptical galaxy M87. In this extreme close-up of M87's core, a jet of particles moving at near light speed streaks toward the upper right corner. At the source of the jet we see a swirling disk of gas. Detailed Hubble observations of this disk's spectrum demonstrate that the disk is rotating at 2 million kilometers per hour (1.2 million miles per hour) around an invisible object weighing two to three billion times more than the Sun. The only thing known to be so heavy and so dark is a massive black hole.

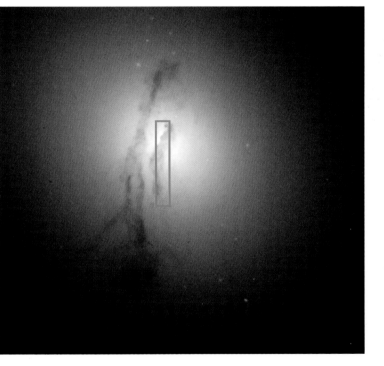
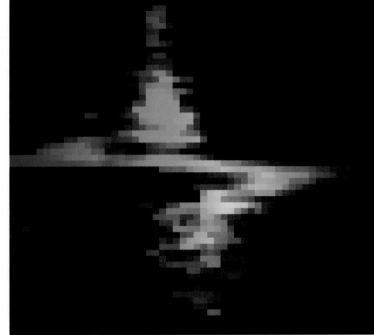

"S" SPELLS SUCCESS

Hubble's light-splitting ability lies at the heart of its talent for detecting black holes. Because of an effect known as the "Doppler shift," light from moving objects subtly changes hue. Colors of objects moving away from us shift slightly to the red end of the spectrum; colors of objects moving toward us shift slightly to the blue end. Careful analysis of the light from matter near a suspected black hole can therefore reveal whether it is orbiting under the influence of a powerful gravitational field.

The Hubble images above show the results of a successful black-hole hunt. At left is the elliptical galaxy M84—the elongated rectangular box outlines the region in which colors were analyzed. To the right is a specialized image showing the spectrum of material at each point along the box, top to bottom. Blue light has been sent to the left side of the image, red light has been sent to the right, and the color shifts have been greatly exaggerated to make them more obvious.

The colorful S shape at the center of this image is the signature of a black hole. The bluish cast of the material just above the center of the box, coupled with the reddish cast just below center, indicates that the light-emitting gas is rapidly orbiting a dark object, almost certainly a black hole. Its orbital speed of 1,400,000 kilometers per hour (880,000 miles per hour) at a distance of twenty-six light-years from the black hole implies that the black hole weighs around 300 million times more than the Sun.

GALAXY DISTANCES

One of the main reasons for building the Hubble Space Telescope was to measure distances to other galaxies as accurately as possible. The galaxy pictured here, a face-on spiral galaxy known as M100, was the first to have its distance measured by Hubble. The close-up Hubble view below shows how M100's feathery spiral arms curl gracefully down to its nucleus. The composite picture opposite above shows how these spiral arms continue out through the rest of the galaxy.

Hubble measures distances to stars in the same way you might estimate the distance to an oncoming car at night. Because you know from experience roughly how much light a car's headlights put out, you can judge about how far away a car is from how bright its headlights appear. In the same vein, astronomers use Hubble to measure the distances to galaxies by measuring the apparent brightness of the galaxy's stars.

Hubble is an unparalleled distance-measuring machine because its superior vision can isolate individual stars in galaxies up to 100 million light-years away, enabling us to measure their brightness. The six-photo sequence opposite below shows Hubble observations of a special "Cepheid variable" star in M100. The time it takes for one of these stars to go from bright to dim to bright again (around forty days in this case) tells us how much light it is putting out. Once we know a star's total light output, we can gauge the distance to the galaxy in which it resides by measuring the star's apparent brightness. Such measurements of this particular Cepheid variable star tell us that M100 sits at a distance of fifty-five to sixty million light-years from the Earth.

EDWIN HUBBLE'S LEGACY

Edwin Hubble himself pioneered the use of Cepheid variable stars to measure distances to galaxies in the 1920s. With the technology of the time, Dr. Hubble could use the Cepheid variable technique only on the closest galaxies and had to resort to less precise distance-measuring methods for galaxies beyond a few million light-years. While crude, these measurements led him to the remarkable insight that the farther a galaxy is, the faster it is moving away from us. This groundbreaking discovery demonstrated that the universe is expanding from a single birth event, and it led to the first estimates of the universe's size and age.

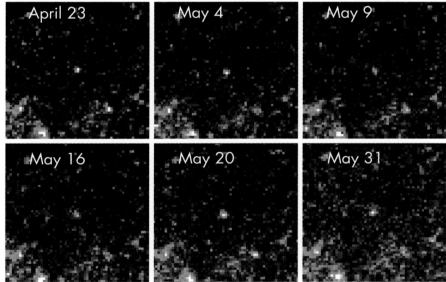

Edwin Hubble died in 1953, and the generation of astronomers following him used the relationship he discovered between a galaxy's distance and its velocity to construct maps of the universe. However, prior to the launch of the Hubble Space Telescope in 1990, the scale of these maps was not well-established. The positions of galaxies relative to one another were well known, but their actual distances were not. By measuring the distances to a few dozen carefully selected galaxies with Hubble, we have finally established the scale of that map to better than 15 percent, providing us with accurate distances to all the other galaxies. Knowing the true distances between the universe's galaxies, we are now better able to calculate how long they have been flying apart. Our best estimates currently place the moment of the Big Bang somewhere between twelve and fourteen billion years ago.

CHAPTER 5
UNIVERSE

A SAMPLE OF THE NORTHERN SKY

The preceding pages display a portion of the Hubble Deep Field, one of the most remarkable astronomical images ever taken. For ten straight days in December 1995, Hubble stared at a single tiny patch of sky and collected all the light it could in over 350 separate exposures. These exposures were then combined to make the Hubble Deep Field, which reveals multitudes of distant galaxies, some too tiny and faint to be detected by any other telescope.

If you live in the Northern Hemisphere and the night is clear, you can observe the Hubble Deep Field for yourself. It lies in the constellation Ursa Major, better known as the Big Dipper, in a thoroughly average region of the sky chosen because it is so typical of the rest of the universe. The blow-out sequence on this page starts with the location of the field relative to the Big Dipper, then zeroes-in on a one-degree square of sky surrounding the field, and finally focuses on the Hubble Deep Field itself. To get a feel for how narrow this patch of sky is, bring a needle with you when you go to observe, hold the needle at arm's length, and look through its eye.

The Hubble Deep Field is located near the Big Dipper for a good reason. Because Hubble circles the Earth every ninety-seven minutes, the Earth usually comes between Hubble and its target once each orbit. Only when Hubble is pointed nearly due north or due south can it stare continually at the sky without the Earth getting in the way. The Hubble Deep Field's location in the far northern part of the sky between the Big Dipper and Polaris, the North Star, enabled Hubble to snap pictures of it, uninterrupted, for a week and a half.

URSA MAJOR

TUCANA

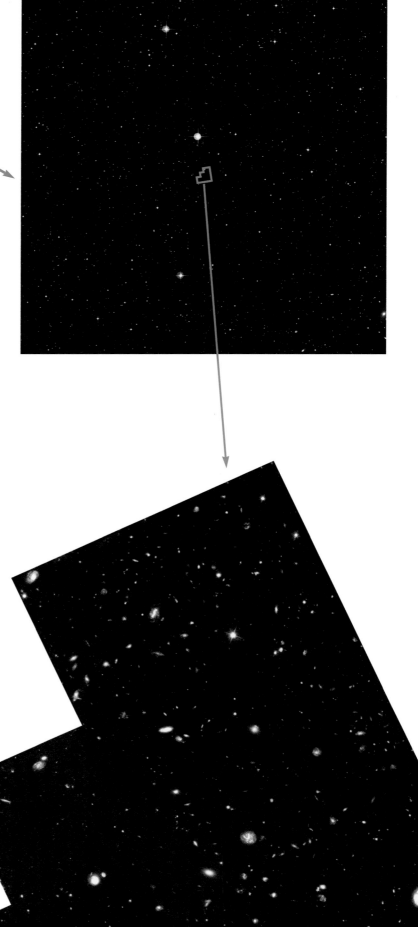

A SAMPLE OF THE SOUTHERN SKY

Telescopes all over the Earth's northern hemisphere have been scrutinizing the site of the Hubble Deep Field ever since its January 1996 release to the public. Never before has such concentrated attention been paid to a single unassuming spot of sky, and the results rank as an unqualified scientific success. In fact, the ensuing cornucopia of discoveries encouraged astronomers to mount a similar campaign in the south. In October of 1998, Hubble once again spent more than a week staring in a single direction, this time gathering light for Hubble Deep Field South. The blow-out sequence on this page shows the full southern field and its location in the constellation Tucana, in the extreme southern reaches of the sky.

Aided by the updated complement of instruments installed on Hubble during its second servicing in 1997, Hubble Deep Field South was even more bountiful than the original. While Hubble's main optical-light camera was recording the image seen here, its newer infrared and optical/ultraviolet cameras were simultaneously staring at neighboring regions of the sky. This comprehensive set of pictures revealed galaxies even fainter than those found in 1995, providing another rich harvest of data on the distant universe.

LAYERS OF HISTORY

Hubble has been called a time machine because it collects light that originated in the distant past, showing us what the universe looked like long ago. If Hubble can really be considered a time machine, then the Hubble Deep Fields represent its deepest excursions back through cosmic history. Light from the nearest stars in these fields has traveled for only a few thousand years, while light from the farthest galaxies began its earthward journey over ten billion years ago, when the universe was only about a billion years old. In between are galaxies spanning over 90 percent of cosmic time, providing clues to the development of the universe that we are now beginning to read.

These pages depict five slices of cosmic history drawn from a section of the northern Hubble Deep Field, shown at the upper right.

◄ OUR BACKYARD: This slice shows the nearest objects in the field—foreground stars in our own galaxy, ranging up to a few tens of thousands of light years away. In comparison with the sweep of cosmic history, the light-travel time from these stars is a brief instant.

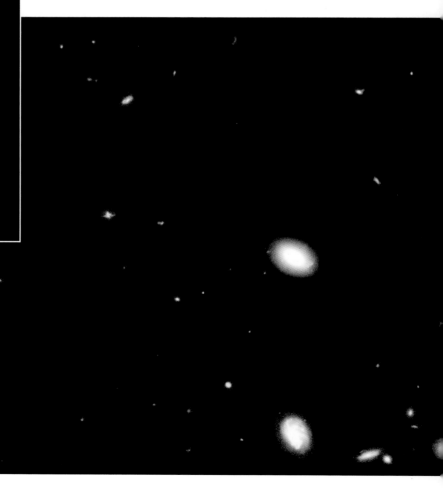

UNIVERSE TODAY: These galaxies are from a time when the universe was about three-quarters of its present age, or nine to fourteen billion years old. This is essentially what the universe looks like today. ▶

◄ HALFWAY BACK: These galaxies show what the universe looked like when it was about one-half of its present age, or six to nine billion years old. For the most part, galaxies still look very much like they do today. The Sun and Earth formed near the end of this era of cosmic history.

YOUNG AND RESTLESS: These galaxies show what the universe looked like when it was about one-quarter of its present age, or three to six billion years old. Some of these galaxies look peculiar because they are temporarily distorted by collisions with other galaxies. Others have the familiar spiral and elliptical shapes of mature present-day galaxies. ▶

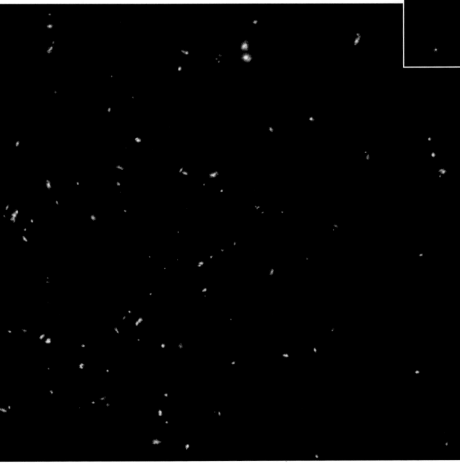

◄ NEAR THE BEGINNING: These are some of the most distant galaxies visible to the Hubble Space Telescope and show what the universe looked like when it was less than three billion years old. Many of the galaxies are small, oddly shaped, and immature.

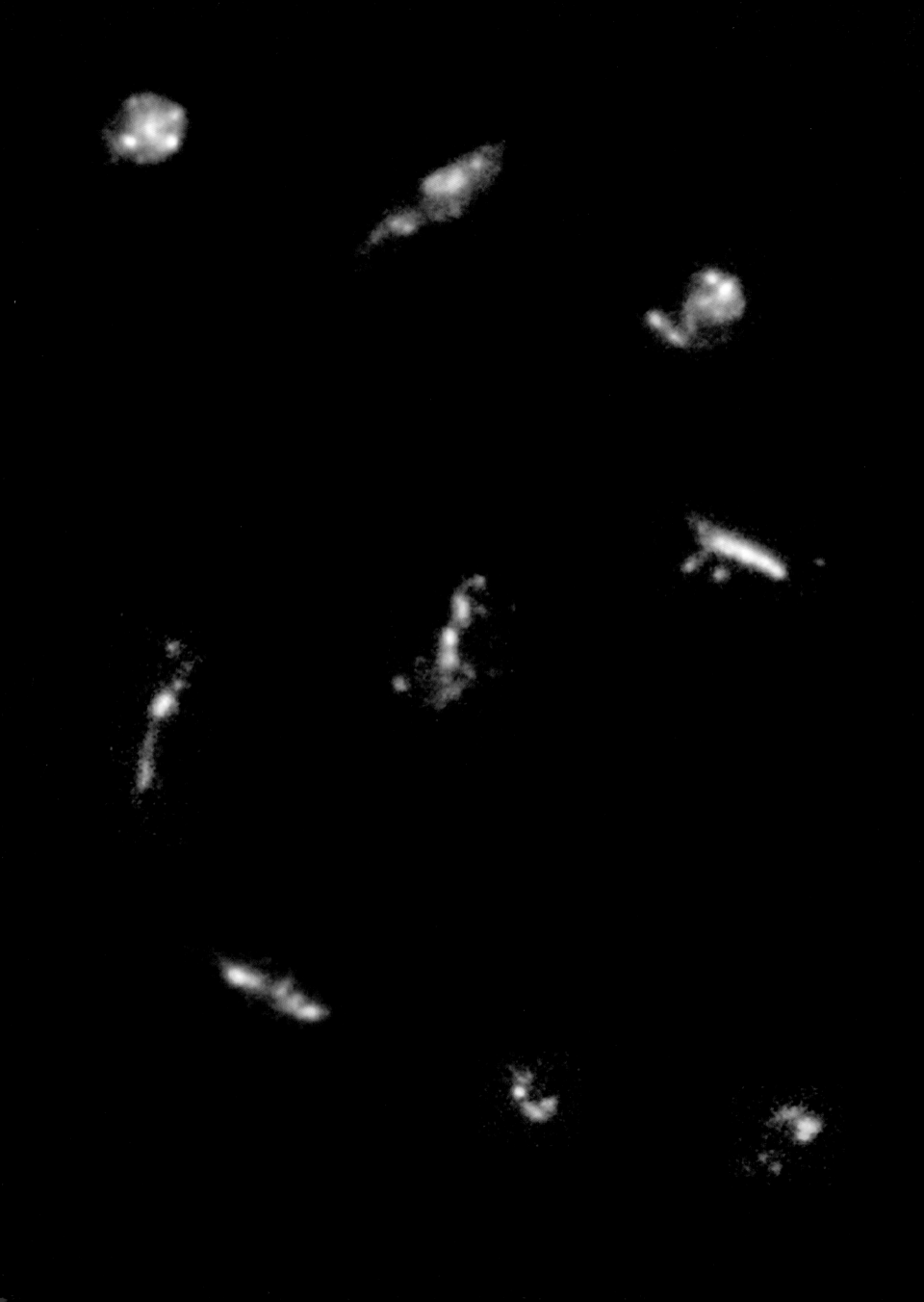

GLIMPSES OF THE VIOLENT PAST

Many of the most distant galaxies Hubble sees appear abnormal. Some of the more unusual ones are shown in the collage on the facing page. These galaxies from the northern Hubble Deep Field span a period from 1.5 to 5.5 billion years after the Big Bang, with the earliest galaxies at the bottom and later ones toward the top. Each of these oddballs probably suffered a disruptive collision shortly before the moment at which we are viewing it, which is not terribly surprising, considering that the universe's galaxies were all much closer together many billions of years ago.

The profusion of strangely shaped galaxies in the early universe supports the most widely held theory of galaxy formation, which asserts that galaxies were created from mergers between smaller blobs of stars and gas. Shortly after the Big Bang, the universe's matter was fairly evenly distributed, but here and there certain regions of space contained slightly more matter than others. From that time on, the rich got richer as the extra gravity of these denser regions drew in matter from surrounding areas too matter-poor to maintain a gravitational grip on their primordial gases. The early blobs of gas that coalesced in this way soon formed stars—so the theory goes—and later collided with one another to form larger pre-galactic units which ultimately assembled themselves into galaxies.

Not even Hubble can see the earliest stages of galaxy formation, but we can test our theories using supercomputers. Starting with some reasonable assumptions about the prevailing conditions in the early universe, we can simulate what ensues according to the known laws of physics. Below are five frames from one such simulation, showing how smaller blobs of matter gradually congregate into a single larger object. In each one, the galaxy under construction looks somewhat distorted, not unlike the Hubble pictures of young, distant galaxies opposite. Agreement between these simulations of the early universe and Hubble's observations is not yet perfect, but the many similarities are quite encouraging.

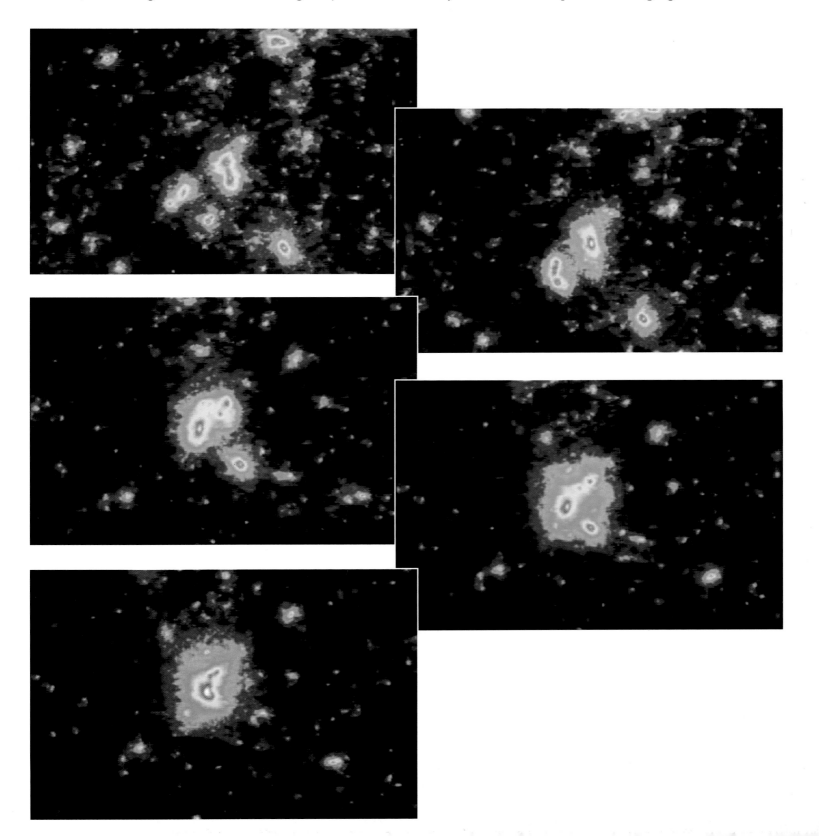

LENSES IN SPACE

Some strange-looking galaxies are the products of collisions, but others are not. A few of the misshapen galaxies we see with Hubble look unusual because of distortions in the fabric of space itself. Massive objects such as clusters of galaxies can bend space in their vicinity, distorting and sometimes magnifying the objects behind them, creating an interesting effect known as gravitational lensing. The picture above shows a particularly striking example. It is centered on a cluster of galaxies about four billion light-years distant. Almost all the yellow galaxies are cluster members, but several blue, ringlike galaxy images (three on the left and one on the right) trace out segments of a circle around the cluster's core. These distorted ovals of light may look like distinct galaxies; amazingly, they are all images of the same distant galaxy lying far beyond the yellow cluster.

In reality, the galaxy responsible for these blue images lies directly behind the cluster, along an axis running from Earth right through the cluster's core. Usually, the only light to reach us from the background galaxy would run directly along this axis; however, distortions of space owing to the cluster's gravity alter the usual situation. Light that would ordinarily pass slightly to one side of the cluster's core gets bent back toward this axis, making it appear to come from a different direction in the sky. In this special example, light from the background galaxy is reaching us from five different directions, producing four distorted images that encircle the cluster core and one faint blue image along the axis passing through the cluster's heart.

OUR ULTIMATE FATE

Images of gravitationally lensed galaxies are fascinating in their own right, but their significance to astronomy runs deeper. Careful analyses of these contorted apparitions are helping to reveal the hidden skeleton of the universe, which consists primarily of matter too dark to be detected by other means. In the case of a galaxy cluster, the amount by which the cluster distorts the image of a background galaxy reflects the mass of the cluster itself— greater distortions imply higher masses. A typical example, the cluster Abell 2218, is shown above. The narrow, arclike galaxy images that trace an intermittent circle around the cluster's center tell us that the vast majority of the cluster's matter is not in the form of visible stars. Because the mysterious unseen matter that fills clusters like this one is so much darker than a cluster's stars, astronomers refer to it as "dark matter."

If the proportions of visible and invisible matter in clusters faithfully reflect the contents of the universe at large, then the masses of clusters are revealing something profound about our cosmic fate. In principle, if the universe contained enough matter, the gravitational pull of each part of the universe on every other part could stop the expansion and reverse it, pulling everything back together in a Big Crunch. From what we now know about the masses of clusters, this outcome seems unlikely. Odds are that the universe's galaxies will continue drifting apart indefinitely into an ever darker, ever more dispersed future.

BEYOND HUBBLE

The Hubble Space Telescope has opened the door to a new era of space astronomy, when fabulous space telescopes of the future will take us all the way back to the births of galaxies, pull back the covers of the stellar cradle, and search for Earthlike worlds with continents, oceans—even life. It represents the first step in NASA's Origins program, a bold attempt to answer some of humanity's oldest questions—where did we come from and are we alone—through astronomical observations.

Hubble Deep Fields have shown us our first glimpses of the earliest galaxies, but even the powerful Hubble Space Telescope cannot guide us back to that exotic time, half a billion years after the Big Bang, when the first generations of stars and the first life-sustaining elements were made. These nascent galaxies are so distant that their light is shifted far to the red end of the spectrum, beyond what our eyes can see. Hubble has instruments that can "see" in the infrared, but its nearness to Earth keeps it warm, meaning that Hubble itself glows with infrared light. Using a warm telescope to look for the birth of a galaxy is like looking for a match in the glow of a kiln!

The way forward is to build cold telescopes that take advantage of the fact that space itself is very cold — this means getting far away from Earth, probably beyond the orbit of the Moon. Within a decade NASA will be building the Next Generation Space Telescope, a venture that will not only detect the youngest galaxies but also allow us to understand in detail how they were assembled and how their births were a step toward the building of planets and life. With a mirror ten times the area of the Hubble's and an orbit around the Sun that takes it far from Earth, the larger telescope will provide superior resolution and much greater sensitivity than any telescope ever built, particularly in the infrared. Its contribution to the study of galaxy birth, galaxy evolution, star and planet formation, and the search for mature planets will be enormous. NASA hopes to launch Hubble's successor in 2009 on a ten-year mission, and even larger space telescopes will soon follow.

Also prominent in NASA's plans for the next two decades is the Terrestrial Planet Finder, a cold space telescope made up of four or more telescopes, each larger than the Hubble. Combining the light from these telescopes, which can be separated by tens, hundreds, perhaps even thousands of meters, will produce vision far sharper than Hubble's. This ultra-precise vision will be crucial in the search for planets around nearby stars, which lie so close to their parent stars that they are lost in the glare. Week- and month-long exposures of these planets' atmospheres may ultimately reveal whether they contain water vapor—which would signal the presence of oceans and lakes—and perhaps even free oxygen, which would point to the probable presence of life.

We stand at the threshold of the greatest era of space exploration in history, thanks to the remarkable success of the Hubble Space Telescope and the vision of those who know that the journey has just begun.

ALAN DRESSLER
CARNEGIE OBSERVATORIES

Simulated deep field from the Next Generation Space Telescope

ON COLOR

Hubble's spectacular images don't emerge from the telescope in vivid color. Each one has been carefully reconstructed from a set of separate black-and-white images in the Hubble Data Archive. Because the telescope's sensitive cameras view celestial objects in only one color of light at a time, Hubble must take a series of pictures to record how an object looks in several different colors of light before a full-color image can be made. Each of the Hubble photographs reproduced in this book uses color in one of the following ways, as noted in the credits on page 62.

TRUE-COLOR IMAGES

Often the goal of a Hubble observation is to capture how an object looks in visible light. Because of the way our eyes work, any visible color can be reproduced by mixing red, green, and blue light in just the right proportions—that's how a television does it. This *true-color* picture of the galaxy NGC 4414 was created in just that way from three separate exposures: one each in red, green, and blue light. Pictures like this one show what the universe would look like if our eyes were as powerful as Hubble.

ENHANCED-COLOR IMAGES

Scientists sometimes use color as a tool to bring out details that would otherwise be too subtle to see. This *enhanced-color* picture of the Cat's Eye Nebula is one example. It was

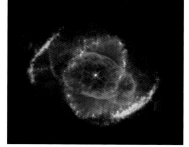

made from three exposures taken of visible light emitted by three different elements. Red represents light from hydrogen, blue represents light from oxygen, and green represents light from ionized nitrogen.

REPRESENTATIVE-COLOR IMAGES

Images taken in infrared light record shades of light that our eyes can't see. In order to visualize these invisible hues, we substitute visible colors

for the infrared ones, creating *representative-color* images like the one of Saturn shown here. Red represents infrared light coming from Saturn's deep cloud layers, green represents light from mid-level clouds, and blue represents light from high-level clouds.

MONOCHROMATIC IMAGES

In some cases, our Hubble images display only one kind of light, allowing us to see differences in brightness from place to place but not differences in color. This *monochromatic* image showing the neighborhood of the central black hole in galaxy M87 is one exam-

ple. Reproducing it in this amber color helps to bring out the contrasts between the ambient light from the galaxy's stars, the disk of gas circling the black hole to the lower left, and the jet of particles shooting to the upper right.

ACKNOWLEDGMENTS

The Hubble Space Telescope is a major scientific accomplishment. Hubble's findings have dramatically altered scientists' understanding of the universe and the public's perception of outer space. Hubble's increasingly available and familiar images offer a visual connection with the cosmos, allowing us to begin to think of it as a less strange and incomprehensible place.

Hubble Space Telescope: New Views of the Universe presents Hubble's contributions to science and education through a unique collaboration between the Smithsonian Institution Traveling Exhibition Service (SITES), one of the nation's foremost museum outreach programs, and the Space Telescope Science Institute (STScI), the scientific organization that operates the Hubble. With the generous support of the National Aeronautics and Space Administration (NASA) Offices of Space Science and Education and Lockheed Martin, we have brought together astrophysicists and exhibit specialists to create two interactive exhibitions for nationwide tour plus complementary educational programs, websites, and this book.

The project was guided by the expertise and support of an incomparable advisory committee, which laid the conceptual groundwork and explicated its main concepts:

- Dr. Russell Hulse, Physicist, Plasma Physics Laboratory, Princeton University, Princeton, New Jersey;
- Dr. Paul H. Knappenberger Jr., President, The Adler Planetarium and Astronomy Museum, Chicago, Illinois;
- Dr. Wendell Mohling, Associate Executive Director, National Science Teachers Association, Arlington, Virginia;
- Dr. Jim O'Leary, Senior Director of Technology and Director of the IMAX Theater, Davis Planetarium, Maryland Science Center, Baltimore;
- Dr. Robert W. Smith, Professor, Department of History and Classics, University of Alberta in Edmonton, Canada;
- Dr. Kathryn Sullivan, Former Astronaut and Director, Center for Science and Industry, Columbus, Ohio; and,
- Dr. Kathryn Thornton, Former Astronaut and Professor of the History of Space Flight, School of Engineering, University of Virginia, Charlottesville.

To move from concept to reality required a project team that combined a vast range of talent, expertise, and commitment. We were fortunate to work with scientists and engineers; machinists and model makers; writers and editors; designers, illustrators, and graphic artists; researchers, teachers, and museum educators; and administrative and technical staffs.

Matou Baillergeau Goodwin, SITES' assistant director of exhibitions, served as project director for *Hubble Space Telescope: New Views of the Universe*, supervising and orchestrating its overall planning, development, and production. Ms. Goodwin's commitment to the highest standards of exhibition design and production was matched only by her tenacity in carrying out the exhibition's educational objectives. She was ably assisted by Meredith Malone, who saw to myriad administrative details. At earlier stages in the project's development, Melissa Doumitt provided valuable assistance as well. Andrea Stevens, SITES' director of external relations, served as senior advisor and directed the development and production of all project publications. Sandra Narva, director of scheduling and exhibitor relations, arranged the exhibition's national tour. Elly Muller, director of public relations, and Lucy Pollio, public relations coordinator, managed all media relations. Jill L. Newmark, registrar, coordinated the details surrounding the acquisition and movement of artifacts, and developed systems for their care and safekeeping during the national tour.

At STScI, Project Manager Patricia Pengra led the scientific team from the very beginning, bringing to the project great innovation, garnering technical and financial support, and unlocking doors to scientific expertise. Without her vigilance and involvement, the exhibition never would have materialized. Carole Rest, deputy project manager, consistently and skillfully resolved innumerable administrative challenges to keep the project running smoothly. Dr. Mark Voit, project scientist, exhibition script developer, and author of this book, patiently taught and shared information. Dr. Carol Christian, head of the Office of Public Outreach provided ongoing support to the realization of this project.

Expertise from the space science community at NASA's Goddard Space Flight Center and Lockheed Martin was critical to exhibition content. David Leckrone, senior project scientist for the Hubble Space Telescope, provided extensive knowledge and support throughout the development of the project. Jim Kelley and his colleagues at Lockheed Martin including Carol DesVoigne, Ron Sheffield, Tom Styczynski, Herb Shodiss, Buzz Bartlett, and Michael Coats shared freely of their direct experience in developing the Hubble Space Telescope. David Phillips, manager of corporate philanthropy, led the effort to secure Lockheed Martin's financial support. In addition to their content generosity, Goddard Space Flight Center, Johnson Space Center, and Lockheed Martin lent original objects and models to the exhibition. Jackson and Tull's contribution of small model telescopes for educational outreach at each host museum is also gratefully acknowledged.

Lee H. Skolnick Architecture + Design Partnership translated extremely complex project concepts and objectives into a lively interactive traveling exhibition. Mr. Skolnick himself served as the exhibition's principal designer with museum projects associate Cynthia E. Smith and designers Scott McQuade and Ostap Rudakevych. Skolnick's dramatic vision was enlightened and interpreted by director of museum services Jo Ann Secor, with Ellen Leerburger, senior interpretive specialist, and Felicia Liss, museum education. The exhibition design team also included Tom Wojciechowski and Harriet Spear of Russell Design Associates; scientific illustrators Don Dixon and Steve Stankiewicz; and photo researcher Audrey Smith O'Malley.

Scientific content and design ideas provided a strong foundation for *Hubble Space Telescope: New Views of the Universe*, but the creation of an experiential environment that would draw visitors into the awesome mysteries of the universe required new presentational technologies. Linda Batwin and Robin Silvestri of Batwin & Robin and Paul Segalla, Gary Letourneau, and their colleagues at ProtoVision delivered that innovation through their exhibition production talents. The astounding scale model of the Hubble telescope that forms the centerpiece of the exhibition was constructed by Barbara Brodsky and Steve Owen of Design Models, Inc. At Chedd-Angier Production Company, Richard Lewis, Kirsten Holmes, Andrew Wilson, and staff created videos and computer interactives.

Richard Maurer wrote the exhibition script, ably translating scientific terminology into engaging text. Dean Trackman edited the exhibition script, developed the brochure's concept, and drafted its text. All of the exhibition's print materials were masterfully designed by Rodney C. Williams and his colleagues at RCW Communication Design Inc. This book would not exist without the leadership and enthusiasm of Eric Himmel, executive editor at Harry N. Abrams, Inc. His personal attention to the content and visual presentation are deeply appreciated. Karen Goldman, Smithsonian contract negotiator, provided valuable assistance with the publishing agreement.

In the acknowledgments above, we have singled out principal members of the project team. Like every undertaking of this magnitude, their work would not have been possible without the valuable contributions of many others. At STScI, we extend special thanks to Greg Bacon, John Bedke, Carl Biagetti, Howard Bond, Stephanie Brown, Sean Cohen, Kathy Cordes, Mark Dickinson, Megan Donahue, Bonnie Eisenhamer, Harry Ferguson, Ann Field, Lisa Frattare, Ginger French, Andy Fruchter, Mauro Giavalisco, John Godfrey, Don Hough, John Isaacs, Stratis Kakadelis, Bill Lane, Zolt Levay, Mario Livio, Piero Madau, Melissa McGrath, Brian McLean, Monique Miskimon, Pat Momberger, Keith Noll, Lydia Paddock, Karen Petro, Mark Postman, Bryan Preston, Cheryl Schmidt, Val Schnader, Peg Stanley, Massimo Stiavelli, John Stoke, Conrad Sturch, Denise Taylor, Terry Teays, Christy Thorndill, Meg Urry, Pat Venanzi, Skip Westphal, Ed Weibe, Eli White, Brad Whitmore, Robert Williams, Monica Wilson, and members of the Hubble Heritage Team.

At NASA Headquarters, special thanks to Dr. Ed Weiler, associate administrator for space science; Dr. Jeffrey Rosendhal, assistant associate administrator for education and outreach; and Dr. Frank Owens, director of the Education Division of the Office of Human Resources and Education.

At NASA Goddard Space Flight Center, we thank Mansoor Ahmed, Wes Alexander, Jimmy Barcus, Nancy Boggess (retired), Lisa Carroll, Frank Cepollina, Mindy DeYarmin, Lee Feinberg, Ann Jenkins, Barbara Lambert, Jack Leibee, Ruth Ann Lewis, Rob Lyle, Gary Moffett, Nancy Roman (retired); Jeff Sandler, Russell Werneth, and John Wood, and staff members of the Scientific Visualization Studio. At the Johnson Space Center in Houston are colleagues who also merit special appreciation: Greg Harbaugh, Mike Gentry, Glen Outz, Ralph Marak, James McBarron, Joyce Berry, and Louis Parker. In addition, we extend gratitude to Tom Arconti, Raytheon; Wendy Freedman, Carnegie Observatory; Stephan Gwyn and Raja Guhathakurta, University of California; Bob Kennicutt, University of Arizona; Robert Kirshner, Harvard-Smithsonian Center for Astrophysics; Bruce Margon, University of Washington; Chris Mihos, Case Western Reserve University; Al Opp (retired), BDM; and Abi Saha, National Optical Astronomy Observatories.

Hubble Space Telescope: New Views of the Universe is an unusually ambitious project with a mission that encompasses broad public outreach and education. Commensurate with its scope, it needed a level of financial support beyond that usually associated with museum exhibits. We were fortunate to have benefited from the extraordinary generosity of the National Aeronautics and Space Administration Offices of Space Science and Education and Lockheed Martin. A grant from the Smithsonian Women's Committee made it possible to produce educational materials for free distribution to teachers. We extend our deepest gratitude to all of these project partners who shared not only our enthusiasm, but also our aspirations for finding a new way of bringing Hubble to Earth.

Anna R. Cohn

Director, Smithsonian Institution Traveling Exhibition Service

CREDITS

NASA's Hubble Space Telescope is operated by the Association of Universities for Research in Astronomy (AURA) at the Space Telescope Science Institute (STScI) in Baltimore, Maryland. All the Hubble images in this book appear courtesy of the National Aeronautics and Space Administration (NASA), STScI/AURA, and the investigators listed below. The single boldface letter associated with each Hubble image identifies the type of color processing used to create it: **T** – true color, **E** – enhanced color, **R** – representative color, **M** – monochromatic. Each process is described on page 60. Many of these spectacular color realizations were created by Zolt Levay and by the Hubble Heritage team who are largely responsible for the beauty of this book.

Front Cover
Ring Nebula / **E** / H. Bond, K. Noll (STScI)

Page 1
Hubble in Earth orbit, 1999 / NASA

Page 5
Edwin Hubble, c. 1922 / Huntington Library

Page 6–7
Hubble Space Telescope, 1997 / NASA

Page 7
Hubble launch, 1990 / NASA
Hubble's third servicing mission, 1999 / NASA

Page 8
Eagle Nebula, ground-based photograph / J. Hester (Arizona State University)
Eagle Nebula / **E** / J. Hester, P. Scowen (Arizona State University)
Star in 30 Doradus, ground-based photograph / G. Meylan (European Southern Observatory)
Star in 30 Doradus / **M** / J. Trauger (Jet Propulsion Laboratory)

Page 9
Helix Nebula, ground-based photograph / J. Bedke (STScI)
Helix Nebula / **E** / C. R. O'Dell (Rice University)
Moon, ground-based photograph / Carnegie Observatories
Moon—Hubble / **M** / J. Caldwell (York University), A. Storrs (STScI)

Pages 10–11
Earth / NASA
Star Cluster M80 / **T** / M. Shara (American Museum of Natural History), F. Ferraro (European Southern Observatory)
All other images credited elsewhere.

Page 12–13
Io and Jupiter in violet and ultraviolet light / **R** / J. Spencer (Lowell Observatory)

Page 14
Venus in ultraviolet light / **R** / L. Esposito (University of Colorado)
Mars / **T** / P. James (University of Toledo), T. Clancy (Space Science Institute), S. Lee (University of Colorado)
Jupiter and Io / **T** / J. Trauger, R. Evans (Jet Propulsion Laboratory)

Page 15
Saturn in infrared light / **R** / E. Karkoschka (University of Arizona)
Uranus in infrared light / **R** / E. Karkoschka (University of Arizona)
Neptune / **E** / L. Sromovsky (University of Wisconsin)
Pluto / **M** / S. A. Stern (Southwest Research Institute), M. Buie (Lowell Observatory)

Page 16
Jupiter with comet fragments / **T** / J. Trauger, R. Evans (Jet Propulsion Laboratory), H. Weaver (Johns Hopkins University), T. E. Smith (STScI)

Page 17
Comet Shoemaker-Levy 9 / **M** / H. Weaver (Johns Hopkins University), T. E. Smith (STScI)
Fragment G impact site / **T** / R. Evans, J. Trauger (Jet Propulsion Laboratory), H. Hammel (Massachusetts Institute of Technology)

Page 18
Mars cyclone / **T** / S. Lee (University of Colorado), J. Bell (Cornell), M. Wolff (Space Science Institute)

Page 19
Martian weather / **T** / S. Lee (University of Colorado), P. James (University of Toledo), M. Wolff (Space Science Institute)
Neptunian weather / **E** / L. Sromovsky (University of Wisconsin)

Page 20
Saturn with auroras / **R** / J. Trauger (Jet Propulsion Laboratory)

Page 21
Io Volcano / **R,M** / J. Spencer (Lowell Observatory)
Jupiter with auroras / **R,T** / J. Clarke (University of Michigan)

Pages 22–23
Keyhole Nebula / **E** / N. Walborn (STScI)

Page 24
Eagle Nebula / **E** / J. Hester, P. Scowen (Arizona State University)
Young stellar disk and jets / **E** / C. Burrows (STScI/European Space Agency)

Page 25
Star Cluster NGC 3603 / **T** / R. Brandner (Caltech), Y. Chu (University of Illinois), E. Grebel (UCO/Lick Observatory)
Betelgeuse in ultraviolet light / **M** / A. Dupree (Smithsonian Astrophysical Observatory), R. Gilliland (STScI)
Planetary Nebula NGC 7027 / **E** / H. Bond (STScI)

Page 26
Orion Nebula / **E** / C. R. O'Dell (Rice University)

Page 27
Young stellar disks in Orion / **E** / C. R. O'Dell (Rice University)
Young stellar jet HH47 / **E** / J. Morse (University of Colorado)

Page 28
Orion star-forming region in infrared light / **R** / R. Thompson (University of Arizona)

Page 29
Orion star-forming region in visible light / **E** / C. R. O'Dell (Rice University)

Egg Nebula in visible light / **M** / R. Sahai, J. Trauger (Jet Propulsion Laboratory)
Egg Nebula in infrared light / **R** / R. Thompson (University of Arizona)

Page 30
Sagittarius star field / **T** / J. Trauger (Jet Propulsion Laboratory)

Page 31
Star cluster NGC 3603 / **T** / R. Brandner (Caltech), Y. Chu (University of Illinois), E. Grebel (UCO/Lick Observatory)
Star cluster NGC 1818 / **T** / R. Elson, R. Sword (Cambridge University), J. Westphal (Caltech)
Star cluster GC1 in M31 / **T** / W. Freedman (Carnegie Observatories), K. Mighell (National Optical Astronomy Observatories), R.M. Rich, J. Neill (Columbia University)

Page 32
Ring Nebula / **E** / H. Bond, K. Noll (STScI)

Page 33
Hourglass Nebula / **E** / R. Sahai, J. Trauger (Jet Propulsion Laboratory)
Cat's Eye Nebula / **E** / J. P. Harrington (University of Maryland), K. Borkowski (North Carolina State University)
Twin-Jet Nebula / **E** / B. Balick (University of Washington), V. Icke (Sterrewacht Leiden), G. Mellema (Stockholm University)

Page 34
Supernova 1987A / **E** / R. Kirshner (Harvard University)

Page 35
Supernova 1987A: Before and After / Anglo-Australian Observatory
Supernova 1987A / **E** / R. Kirshner (Harvard University)
Spectrum of Supernova 1987A / **T** / G. Sonneborn (NASA)

Pages 36–37
Galaxies NGC 2207 and IC 2163 / **T** / D. Elmegreen (Vassar College)

Page 38
Galaxy NGC 4414 / **T** / W. Freedman (Carnegie Observatories)
Galaxy NGC 7742 / **T** / J. Westphal (Caltech)

Page 39
Polar-ring galaxy NGC 4650A / **T** / J. Gallagher, L. Sparke (University of Wisconsin), A. Kinney (NASA), L. Matthews (National Radio Astronomy Observatory)
Hickson Compact Group 87 / **T** / J. Charlton (Pennsylvania State University), S. Hunsberger (Lowell Observatory)

Page 40
Antennae Galaxies / **T** / B. Whitmore (STScI)
Antennae Galaxies, ground-based photograph / F. Schweizer (Carnegie Institution)

Page 41
Cartwheel Galaxy / **T** / K. Borne (Hughes STX)
Cartwheel Galaxy simulation / C. Mihos (Case Western Reserve University), G. Bacon (STScI)

Page 42
Galaxy Centaurus A / **T** / E. Schreier (STScI)

Page 43
Centaurus A, ground-based photograph / National Optical Astronomy Observatories
Centaurus A nucleus in infrared light / **R** / E. Schreier (STScI)

Page 44
Nucleus of Galaxy M87
H. Ford (Johns Hopkins University)

Page 45
Galaxy M84 / **M** / G. Bower, R.Green (National Optical Astronomy Observatories)
Spectrum of M84 nucleus / **M** / G. Bower, R. Green (National Optical Astronomy Observatories)

Page 46
Galaxy M100 / **T** / J. Trauger (Jet Propulusion Labaratory)

Page 47
M100, ground-based photograph / Palomar Observatory
Cepheid variable star in M100 / **M** / W. Freedman (Carnegie Observatories)

Page 48–49
Hubble Deep Field / **T** / R. Williams (STScI)

Page 50
Hubble Deep Field / **T** / R. Williams (STScI)
Region around Hubble Deep Field / Palomar Observatory

Page 51
Hubble Deep Field South / **T** / R. Williams (STScI)
Region around Hubble Deep Field South / Anglo-Australian Observatory

Pages 52–53
Layers of Hubble Deep Field / **T** / S. Gwyn (University of Victoria)

Page 54
Distant galaxies in Hubble Deep Field / **T** / H. Ferguson (STScI), Y. Park (Johns Hopkins University)

Page 55
Simulation of galaxy formation / E. Bertschinger (Massachusetts Institute of Technology)

Page 56
Galaxy Cluster 0024–1654 / **T** / W. Colley (Harvard-Smithsonian Center for Astrophysics), E. Turner (Princeton University), J. A. Tyson (Lucent Technologies)

Page 57
Galaxy Cluster Abell 2218 / **T** / A. Fruchter (STScI)

Page 58
Simulated Deep Field / M. Im (Johns Hopkins University)

Back Cover
Jupiter with comet fragments / **T** / J. Trauger, R. Evans (Jet Propulsion Laboratory), H. Weaver (Johns Hopkins University), T. E. Smith (STScI)

Published on the occasion of a traveling exhibition, *Hubble Space Telescope: New Views of the Universe*, organized by the Smithsonian Institution Traveling Exhibition Service and the Space Telescope Science Institute, operated for the National Aeronautics and Space Administration by the Association of Universities for Research in Astronomy, Inc. The exhibition and its educational programs have been made possible through the generous support of the National Aeronautics and Space Administration Offices of Space Science and Education and Lockheed Martin.

Editor: ERIC HIMMEL
Designer: RAYMOND P. HOOPER

Library of Congress Cataloging-in-Publication Data

Voit, Mark
    Hubble space telescope : new views of the universe / by Mark Voit.
        p.   cm.
    ISBN 0–8109–2923–6 (pbk.)/ ISBN 0–8109–3447–7 (book club: hc)
    1. Universe—Pictorial works.   2. Hubble Space Telescope (Spacecraft).   I. Title
    QB500.25.V65   2000
    520'.22'2—dc21   00–029316

Copyright © 2000 Smithsonian Institution and the Association of Universities for Research in Astronomy, Inc.

Published in 2000 by Harry N. Abrams, Incorporated, New York
All rights reserved. No part of the contents of this book may be reproduced without the written permission of the publisher

Printed and bound in Hong Kong

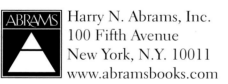 Harry N. Abrams, Inc.
100 Fifth Avenue
New York, N.Y. 10011
www.abramsbooks.com

10 9 8 7 6 5 4 3